Design Principles and Analysis
of Thin Concrete Shells,
Domes and Folders

Design Principles and Analysis of Thin Concrete Shells, Domes and Folders

Iakov Iskhakov and Yuri Ribakov

Ariel University, Israel

CRC Press
Taylor & Francis Group
Boca Raton London New York

CRC Press is an imprint of the
Taylor & Francis Group, an **informa** business

A SCIENCE PUBLISHERS BOOK

CRC Press
Taylor & Francis Group
6000 Broken Sound Parkway NW, Suite 300
Boca Raton, FL 33487-2742

First issued in paperback 2019

© 2016 by Taylor & Francis Group, LLC
CRC Press is an imprint of Taylor & Francis Group, an Informa business

No claim to original U.S. Government works

ISBN-13: 978-1-4987-2664-1 (hbk)
ISBN-13: 978-0-367-37721-2 (pbk)

Visit the Taylor & Francis Web site at
http://www.taylorandfrancis.com

and the CRC Press Web site at
http://www.crcpress.com

Preface

One of the main goals of a good and effective structural design is to decrease, as far as possible, the self-weight of structures, because they must carry the service load. This is especially important for reinforced concrete (RC) structures, as the self-weight of the material is substantial. For RC structures it is furthermore important that the whole structure or most of the structural elements are under compression with small eccentricities. Continuous spatial concrete structures satisfy the above mentioned requirements. It is shown in this book that a span of a spatial structure is practically independent of its thickness and is a function of its geometry. It is also important to define, which structure can be called a spatial one. Such a definition is given in the book and based on this definition, five types of spatial concrete structures were selected: translation shells with positive Gaussian curvature, long convex cylindrical shells, hyperbolic paraboloid shells, domes and long folders.

To demonstrate the complex research, results of experimental, analytical and numerical evaluation of a real RC dome are presented and discussed. The book is suitable for structural engineers, students, researchers and faculty members at universities.

Contents

Introduction

Spatial reinforced concrete structures were widely used during the second half of last century, as they enabled to effectively cover larger spans in both directions without the placement of interior columns. The advantages of these structures are due to the single- or double-curvature of their surface (yielding essential decrease in bending moments and shear forces) and economic efficiency (as the minimum thickness of the structure is usually a few centimeters only). In recent decades, additional benefits related to numerical modeling, advances in formwork and new concrete technologies (for example, the use of steel fibers), opened new possibilities for using spatial structures. Essentially, one requires to gain a deeper analysis of buckling problems, load bearing capacity in ultimate limit state and so on (for example, considering physical and/or geometric non-linearity in numerical analysis).

One of the main purposes of a correct design is to decrease the self-weight of the structure as much as possible, because the role of the structure is to carry the live load. This is particularly important in reinforced concrete structures, because the self-weight of the material is rather high. This issue is even more important in regions with high seismic activity as the seismic forces are proportional to the mass of the structure. It is very important that the whole reinforced concrete (RC) structure or its main part should be in pure compression or in compression with small eccentricity. Spatial RC structures completely satisfy all these requirements. A span of a spatial structure is independent of its thickness, as in plane ones, but depends on their geometry as it is proved below.

This monograph deals with the theory of surfaces and corresponding elements of differential geometry, analysis, design and investigation of spatial structures with continuous surface that are most used in newly constructed shells (not reticulated three-dimensional structures!). It is focused on analytical simplified calculation methods of shells that are very important at initial design stages, for example for selecting optimal kind of shell for given materials and loads. The main criterion for selecting proper material is distribution of internal forces. For convex compressed shells concrete is preferred; if the element is in tension, steel or fiberglass are more suitable. Using timber shells is efficient in cases when the dominant load is due to its self-weight.

For the first time a definition of a shell structure is given and the difference between shells and other kinds of structures (linear or plane ones) is emphasized. Additionally, in the frame of this book, the problem of buckling in case of simultaneous physical and geometrical shell non-linearity is considered as one of possible limit states of the structure. Recommendations for selecting initial dimensions for load bearing elements of a spatial structure (thickness, rise, edge elements section, ties, openings, etc.) are given as well. These recommendations are very important during the early design stages of the shell.

Computer-based design that is widely used in recent decades provides information on how the structure behaves, but do not explain the reason of such behavior. Therefore, results of such calculations cannot be used for other structures and in any other case new analysis should be carried out. Analytical methods are more universal and visual. They allow obtaining the critical load at early design stages. Therefore, the book is focused on analytical and experimental approaches.

The shells discussed in this work are mainly covering structures subjected to self-weight loads. Therefore, these shells are thin-walled and shallow, significantly decreasing their mass and making the construction easier. Most shells described in this book are convex at least in one direction. Therefore, they are mainly in compression, which is optimal for applying thin-walled RC shells.

The book includes numerical examples on surface geometry and calculation of many various typical shells, based on analytical methods. Different types of shell structures and their edge elements are presented, including those, constructed in the recent years.

Let's assume a quadratic plate with thickness h_1 and span l_1 subjected to a uniformly distributed load g_1 due to its self-weight. The plate is simply supported by beams along its perimeter (Figure 1a). The strength of the material is f_{m1} and its specific gravity is γ_1. It is possible to calculate the load bearing capacity of the plate according to the following known dependence:

$$m = g_1 \, l_1^2 / 24 \text{ kN m/m}$$

The maximum stress is

$$\sigma_{m1} = m/W = 4\, g_1\, l_1^2/h_1^2 = \gamma_1\, l_1^2/(4\, h_1^2); \; g_1/h_1 = \gamma_1 \text{ kg/m}^3$$

and when the stress equals to the strength

$$l_1 = 2\, (f_{m1}/\gamma_1)^{0.5}\, h_1^{0.5} \tag{1}$$

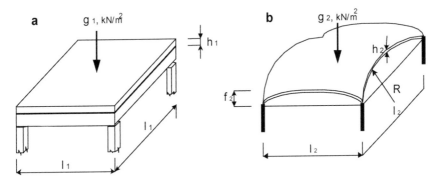

Figure 1. A planar structure-plate (a) and a spatial structure-shell (b).

Let consider a rectangular convex shell that has a form of elliptic paraboloid with parameters h_2, l_2, γ_2, f_{m2}. The shell rise is f_2 the load due to the self-weight is g_2 (see Figure 1b). The shell is supported by edge elements (diaphragms) that take the spreading forces. According to the shells theory (that will be explained in details later), the maximum compression force in the center of the shell (shell vertex) is

$$N = 0.5\, g_2\, R = g_2\, l_2^2/(16 f_2)\ \text{kN/m} = \gamma_2\, h_2\, l_2^2/(16 f_2)\ \text{kN/m}$$

where the radius of curvature of the shell

$$R = l_2^2/(8 f_2)$$

The maximal strength

$$\sigma_{m2} = N/h = \gamma_2\, l_2^2/(16 f_2)$$

Therefore (when the stress is equal to strength)

$$l_2 = 4\, (f_{m2}/\gamma_2)^{0.5}\, f_2^{0.5} \tag{2}$$

From comparison of expressions (1) and (2), it is evident that the plate span depends on its thickness, whereas the span of the shell is a function of its rise only. Hence, it is logically to cover big spans by spatial structures. The border between a span of planar and spatial structure is about 18 m.

Finally, just one more example. The ratio between the egg shell thickness (about 0.2 mm) and its diameter (about 50 mm) is 1:250. A ratio for typical shells is about 1:400 (see a more detailed information in Chapter 3). Therefore, shells are indeed thin-walled and lightweight structures!

1

General Concepts of Differential Geometry and Surface Theory

1.1 Main concepts of differential geometry

Spatial structures are three-dimensional ones and to determine them it is necessary to operate by mathematical terms of surface theory and differential geometry. This chapter deals with the above mentioned aspects, but first basic concepts in planes curves and local elements of a curve are presented, and the definition of curvature is also given.

1.1.1 Plane curves in different systems of coordinates

As known, it is possible to describe any curve analytically, using one of the following coordinate systems:

- orthogonal Cartesian coordinate system, or
- polar coordinate system.

In the first case an appropriate equation in an evident form is

$$y = f(x) \qquad (1.1.1)$$

where x and y are coordinates of a point, belonging to the curved line.

For example, an equation of a sinusoidal curve is:

$$y = \sin x$$

If a non-evident form is used, then

$$f(x, y) = 0 \qquad (1.1.2)$$

For example, for a circle line

$$x^2 + y^2 = r^2$$

If a parametric form is used, then

$$x = x\,(t),\, y = y\,(t) \tag{1.1.3}$$

where t is a parameter in orthogonal coordinates.

For example, for a circle line (Figure 1.1.1a)

$$x = r\cos t, \quad y = r\sin t$$

If a polar coordinate system is used, then (Figure 1.1.1b)

$$\rho = f\,(\varphi) \tag{1.1.4}$$

where ρ and φ are polar radius and angle, respectively.

For example, for Archimedes' spiral

$$\rho = a\,\varphi$$

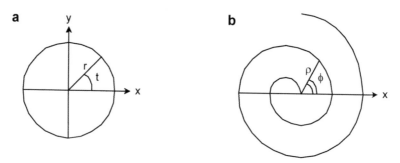

Figure 1.1.1. A scheme for parametric equation of a circle (a); Archimedes' spiral (b).

1.1.2 Local elements of a curve

It is possible to define a differential of a curve at point M, ds, by segment MN if its length is close to zero. This differential can be expressed according to Eqs. (1.1.1–1.1.3) as follows (Figure 1.1.2a):

$$\Delta s \approx ds = \sqrt{1 + (\frac{d\,y}{d\,x})^2}\ dx = \sqrt{x'^2 + y'^2}\ dt = \sqrt{\rho^2 + \rho'^2}\ d\varphi \tag{1.1.5}$$

For example:

$$y = \sin x, \; ds = (1 + \cos^2 x)^{1/2} \, dx;$$

$$x = t^2, \quad y = t^3, \; ds = t \, (4 + 9 \, t^2)^{1/2} \, dt;$$

$$\rho = a \, \varphi, \quad ds = a \, (1 + \varphi^2)^{1/2} \, d\varphi$$

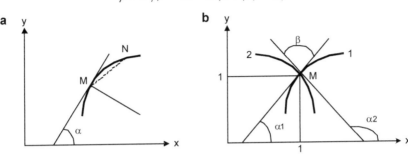

a

b

Figure 1.1.2. A tangent to a curved line (a) and an angle between two curved lines (b).

A tangent to a curved line at point M is the limit location of a straight line MN, when N → M. A normal to the tangent at point M is normal to the curved line. The slope of the tangent is defined by angle α. For example,

$$y = \sin x, \quad \tan \alpha = dy/dx = \cos x \, dx/dx = \cos x, \quad \alpha = \text{arc tan } (\cos x);$$

$$x = t^2, \quad y = t^3, \quad \tan \alpha = dy \, (t)/dx \, (t) = 3 \, t^2/2 \, t = 3 \, t/2, \quad \alpha = \text{arc tan } (3 \, t/2);$$

$$\rho = a \, \phi, \quad \tan \alpha = \rho/(d\rho/d\phi) = a \, \phi/a = \phi, \quad \alpha = \text{arc tan } \phi$$

An angle β between two curves at their intersection point M is equal to the angle between the tangents to those lines at this point. For example, Let's find the angle β between two curves, given by the following equation, at the intersection point M (1, 1):

$$y_1 = x_1^{1/2}, \quad y_2 = x_2^2,$$

$$\tan \alpha_1 = dy_1/dx_1 = 0.5 \, x_1^{-1/2}, \quad \tan \alpha_2 = dy_2/dx_2 = 2 \, x_2$$

At the intersection point $x_1 = x_2 = 1$, therefore

$$\tan \beta = \frac{\tan \alpha_2 - \tan \alpha_1}{1 + \tan \alpha_2 \, \tan \alpha_1} = \frac{2 - 0.5}{1 + 2 \cdot 0.5} = 0.75$$

$$\beta = \text{arc tan } 0.75 = 36.87°$$

1.1.3 Concave and convex curves bending point

If at any point M that belongs to a curve, defined by equation $y = f(x)$, the second derivative is positive:

$$y'' > 0 \tag{1.1.6}$$

then the curve is concave upward. Conversely, if

$$y'' < 0 \tag{1.1.7}$$

the curve is convex upward. When

$$y'' = 0 \tag{1.1.8}$$

the curve bends at point M, i.e., M is the bending point.

For example, for a third order parabola $y = x^3$, then $y'' = 6x^2$. In this case the curve convexity depends on x (Figure 1.1.3). If point M is located in the positive half of the horizontal axis, i.e., $x > 0$, then $y'' > 0$ and the curve is concave and wise versa, if $x < 0$ then $y'' < 0$ the curve is convex.

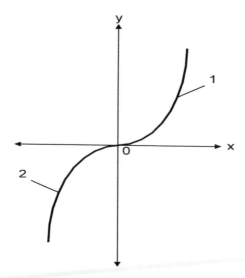

Figure 1.1.3. Branches of a cubic parabola:
1—concave upward; 2—convex upward; 0—bending point of the curve.

1.1.4 Curvature and radius of curvature

The curvature k of a curve at any point M is a limit ratio between the angle $(\alpha_N - \alpha_M)$ and the curve's length ds when MN $\rightarrow 0$ (Figure 1.1.4):

$$k = \lim_{ds \to 0} \left[(\alpha_N - \alpha_M)/ds = d\alpha/ds = d^2y/dx^2 \right] \tag{1.1.9}$$

$$(\alpha = \alpha_N - \alpha_M \approx \tan \alpha = dy/dx; \; d\alpha/ds = d^2y/dx^2)$$

where α_N and α_M are angles between the X axis and tangent lines at points M and N respectively.

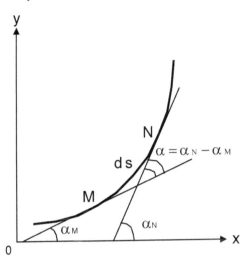

Figure 1.1.4. Definition of curve's curvature at point M.

The curvature k has a sign that depends on the left side part in Eq. (1.1.9). But this sign is not always required. In those cases an absolute value of curvature is used. Radius of curvature, R, at point M is an inverse of curvature k, i.e.,

$$R = 1/k \tag{1.1.10}$$

For a circle line with radius r the centers of curvature of all points that belong to the circle are located at the center of the circle line. The beginning of R is the center of curvature of M:

$$R = r = \text{const}, \quad k = 1/r = \text{const} \tag{1.1.11}$$

For any point, belonging to a straight line

$$R = \infty = \text{const}, \quad k = 0 = \text{const} \tag{1.1.12}$$

For a second order parabola (Figure 1.1.5)

$$y(x) = 4f(0.5 - x/L)^2$$

$$d^2y/dx^2 = k = 8f/L^2 = \text{const}, \quad R = L^2/8f = \text{const} \qquad (1.1.13)$$

However, the curvature centers for different points that belong to the parabola are not at the same point, like for circle—they are located at a line that is "parallel" to the given parabola.

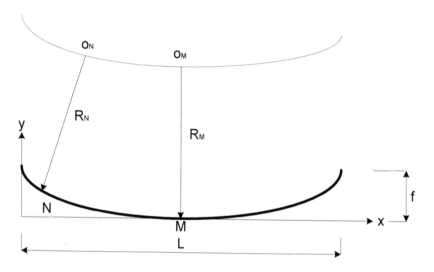

Figure 1.1.5. Location of curvature centers for a second order parabola.

For other types of curves (except circle, second order parabola and straight line) the curvature changes from point to point, i.e., it is not constant.

1.1.5 Spatial curves

A spatial curve can be determined analytically by one of the following methods:

- intersection between two surfaces given by equations

$$f(x, y, z) = 0, \quad F(x, y, z) = 0 \qquad (1.1.14)$$

- using a parametric method (t is some parameter, e.g., $t = s$, where s is the length of the spatial curve)

$$x = x(t), \quad y = y(t), \quad z = z(t) \qquad (1.1.15)$$

The length of a spatial curve, s, between points A and M can be calculated as follows:

$$s = \int_{t_A}^{t_M} \sqrt{\left(\frac{dx}{dt}\right)^2 + \left(\frac{dy}{dt}\right)^2 + \left(\frac{dz}{dt}\right)^2}\ dt \qquad (1.1.16)$$

The curvature of a spatial curve, k, at point M is a number that characterizes the deviation of the curve from the tangent straight line (Figure 1.1.6a):

$$k = \lim_{MN \to 0} \left|\frac{\Delta t}{M\ N}\right| = \left|\frac{dt}{ds}\right| \qquad (1.1.17)$$

a

b

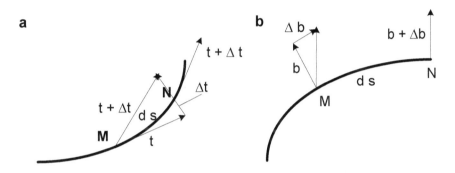

Figure 1.1.6. Curvature (a) and bending (b) of a spatial curve at point M: Δt—deviation of the curve from the straight line; Δb—deviation of the spatial curve from the plane one.

Bending of a spatial curve T at point M is a numerical value that characterizes deviation of the curve from a plane one:

$$T = \lim_{MN \to 0} \left|\frac{\Delta b}{M\ N}\right| = \left|\frac{db}{ds}\right| \qquad (1.1.18)$$

1.2 Concepts in the theory of surfaces

1.2.1 Surface equation

A spatial surface geometry can be described by equations in the following forms:

- evident form

$$z = f(x, y) \qquad (1.2.1)$$

- non-evident form

$$f(x, y, z) = 0 \qquad (1.2.2)$$

- parametric form

$$x = x\,(u, v), \quad y = y\,(u, v), \quad z = z\,(u, v) \qquad (1.2.3)$$

For example, for a spherical shape

$$z = (r^2 - x^2 - y^2)^{1/2}$$

$$x^2 + y^2 + z^2 - r^2 = 0$$

$$x = r \cos u \sin v, \quad y = r \sin u \sin v, \quad z = r \cos v \qquad (1.2.3a)$$

as shown in Figure 1.2.4b.

1.2.2 Geometry of continuous surfaces

Generally, just two types of surfaces are used for design of continuous spatial structures: revolution and translation surfaces. This chapter deals with those two types of surfaces.

A revolution surface is obtained by rotating any plane shape around an axis (axis of rotation). An example of revolution surfaces is an onion-shape shell (Figure 1.2.1a). A translation surface is obtained as result of transition of a creating line along two track lines (for example, saddle-shape surface shown in Figure 1.2.1b).

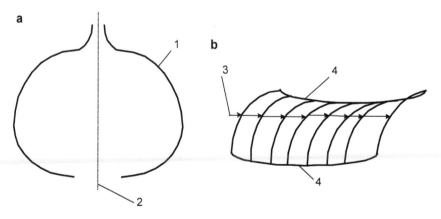

Figure 1.2.1. Revolution onion-shape surface (a) and translation saddle-shape surface (b): 1—plane line; 2—rotation axis; 3—creating line; 4—tracks for creating line.

1.2.3 Equations of revolution surfaces

An equation of a spherical surface (Figure 1.2.2a) with a center, located at the origin of spatial coordinates system, is:

$$z = (R^2 - x^2 - y^2)^{1/2} \qquad (1.2.4)$$

An equation of revolution paraboloid, created by rotating a second order parabola, $z = x^2$, around axis z (Figure 1.2.2b) is:

$$z = f(x^2 + y^2)/r^2 \qquad (1.2.5)$$

where f is the shell rise and r is a radius of revolution.

An equation of revolution ellipsoid (similar to the shape of the Earth) is:

$$\frac{x^2}{a^2} + \frac{y^2}{b^2} + \frac{z^2}{c^2} - 1 = 0 \qquad (1.2.6)$$

where a, b and c are halves of the ellipsoid axes in direction of x, y and z respectively.

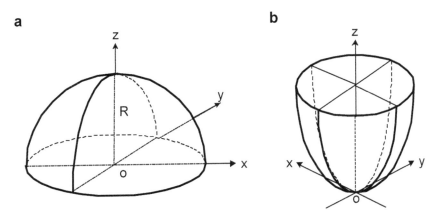

Figure 1.2.2. Spherical surface (a) and surface of revolution paraboloid (b).

1.2.4 Equations of translation surfaces

An equation of an elliptic paraboloid on rectangular basis, $l_1 \times l_2$ (Figure 1.2.3a), is:

$$z = f_1 \frac{x^2}{a^2} + f_2 \frac{y^2}{b^2} \qquad (1.2.7)$$

where a and b are halves of spans l_1 and l_2, respectively;

f_1 and f_2 are rises of curves in l_1 and l_2 directions, respectively.

An equation of a saddle-shape hyperbolic paraboloid (hypar), shown in Figure 1.2.3b, is:

$$z = f_1 \frac{x^2}{a^2} - f_2 \frac{y^2}{b^2} \tag{1.2.8}$$

and for a simple linear hypar (Figure 1.2.3c)

$$z = \frac{f}{a\,b}\, x\, y \tag{1.2.9}$$

where f is the hog of the hypar.

Such types of surfaces are called surfaces with double curvature.

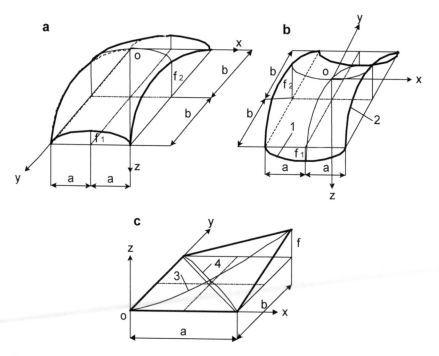

Figure 1.2.3. Shapes of surfaces with double curvature:

(a) elliptic paraboloid; (b) saddle-shaped hyperbolic paraboloid; (c) linear-shape paraboloid (simple hypar); 1 and 2—convexity and concavity directions of the surface parallel to its sides; 3 and 4—convexity and concavity directions of the surface perpendicular to its diagonals.

1.2.5 Surface curved coordinate system

Let's assume a surface, given in a parametric form (see Eq. 1.2.3), or in a form of a vector is:

$$r = r(u, v); \qquad r = x(u, v)\, i + y(u, v)\, j + z(u, v)\, k \qquad (1.2.10)$$

If one of parameters is defined as $v = v_0$ and the second, u, varies then point $r(x, y, z)$ describes a curve $r = r(u, v_0)$ on the given surface. Varying the constant value of v_0 ($v = v_1$, $v = v_2$ ...), leads to a system of curves on the surface. Similarly, if $u = u_0$, it is possible to obtain a curve $r = r(u_0, v)$ that also belongs to the surface. Different constant values of parameter $u = u_1$, $u = u_2$, etc. enables to obtain another system of curves on the surface. In such a way a net is created on the surface. This net forms a system of curved coordinates. Two selected numbers $u = u_i$, $v = v_k$ define the curved coordinates (or Gaussian coordinates) of point M that belongs to the surface (Figure 1.2.4a).

If a surface is given in an evident form (see Eq. 1.2.1), then its intersection with planes $x =$ const, $y =$ const are the coordinates' lines. Each equation, relating to the above mentioned coordinates, i.e., F (u, v) or $u = u(t)$, $v = v(t)$, defines some curve on the surface. For example, in an equation of a sphere (see Eq. 1.2.3a) u is the geometric longitude of point M (u is equal to the angle *pox*), v is a polar value of the point (v is the angle M*oz*). Lines v are meridians AMB and lines u are the parallels (Figure 1.2.4b).

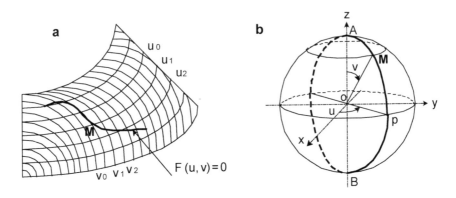

Figure 1.2.4. Curved coordinates of a surface in general case (a) and of a sphere (b).

1.3 Surface curvature

1.3.1 Differential of a curve

Let's assume that $M(u, v)$ is any point at a surface, given in parametric form, or in a form of a vector, and point N $(u + du, v + dv)$ is near to M and belongs to the same surface. Then, the length of segment MN on the surface is a differential of a curve (sometimes it is also called a linear element of the surface or a first quadratic shape of the surface):

$$G_I = ds^2 = E\ du^2 + 2\ F\ du\ dv + G\ dv^2 \tag{1.3.1}$$

$$ds = (E\ du^2 + 2\ F\ du\ dv + G\ dv^2)^{0.5}$$

where E, F, G are the first quadratic shape coefficients of the surface;

$$E = r_1^2 = \left(\frac{\partial x}{\partial u}\right)^2 + \left(\frac{\partial y}{\partial u}\right)^2 + \left(\frac{\partial z}{\partial u}\right)^2, G = r_2^2 = \left(\frac{\partial x}{\partial v}\right)^2 + \left(\frac{\partial y}{\partial v}\right)^2 + \left(\frac{\partial z}{\partial v}\right)^2,$$

$$F = r_1\, r_2 = \frac{\partial x}{\partial u}\frac{\partial x}{\partial v} + \frac{\partial y}{\partial u}\frac{\partial y}{\partial v} + \frac{\partial z}{\partial u}\frac{\partial z}{\partial v} \tag{1.3.2}$$

For example, for a sphere, defined as a vector (i.e., in a non-evident form)

$$r = a\ (\cos u \sin v\ \boldsymbol{i} + \sin u \sin v\ \boldsymbol{j} + \cos v\ \boldsymbol{k}),$$

$$E = a^2 \sin^2 v,\ F = 0,\ G = a^2,\ G_I = a^2\ (\sin^2 v\ du^2 + dv^2) \tag{1.3.3}$$

(proof: $u = 0$ is constant, v is variable, $r = a\ (\sin v + 1)$, $r_1 = a \sin v\ \boldsymbol{i}$, $r_2 = a\ \boldsymbol{j}$).
If any surface is given in an evident form, i.e., $z = f(x, y)$, then

$$E = 1 + p^2, \qquad F = p\, q, \qquad G = 1 + q^2 \tag{1.3.4}$$

$$p = \frac{\partial z}{\partial x}, \qquad q = \frac{\partial z}{\partial y} \tag{1.3.5}$$

1.3.2 Length of a curved line and surface area

The length of a curved line $u = u\ (t)$, $v = v\ (t)$ on a surface for $t_0 \le t \le t_1$ can be found, using the following expression:

$$L = \int_{t_0}^{t_1} ds = \int_{t_0}^{t_1} \sqrt{E\left(\frac{\partial u}{\partial t}\right)^2 + 2\ F\ \frac{\partial u}{\partial t}\frac{\partial v}{\partial t} + G\left(\frac{\partial v}{\partial t}\right)^2}\ dt \tag{1.3.6a}$$

In a particular case, when a curved line on a surface is given in an evident form $z = f(x, y)$, its length can be found as follows:

$$L = \sqrt{E\,(dx)^2 + 2\,F\,dx\,dy + G\,(dy)^2} \qquad (1.3.6b)$$

An area of surface S, defined by a curve, given by a double integral of the surface differential is:

$$S = \iint_S dS \qquad (1.3.7)$$

$$dS = \sqrt{E\,G - F^2}\;du\,dv \qquad (1.3.8)$$

1.3.3 The surface curvature

Let's assume that point M belongs to a spatial curve G on surface S. Let ρ be the curvature radius of G at point M. Let's also assume that M belongs also to a plane line C that is an intersection of plane P with the surface (Figure 1.3.1a).

Definitions

a) The curvature radius ρ in plane P of the spatial line G is equal to the curvature radius of curve C at point M;
b) The curvature radius of the planar line C is equal to (Figure 1.3.1b)

$$\rho = R \cos (n, N) \qquad (1.3.9)$$

where R is the curvature radius of the intersection line, C_{norm}, between the surface S and a plane, perpendicular to the surface (definition of the perpendicular plane will be given below);

N and n are directions of lines that are perpendicular to the intersection line at point M and belong to planes that include lines C and C_{norm} (Magnel theorem).

The sign of R in Eq. (1.3.9) is positive (negative) if the radius is in the concave (convex) direction of line C_{norm} (in Figure 1.3.1 the radius direction is positive);

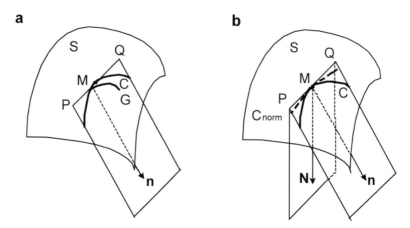

Figure 1.3.1. Plane line C as intersection between the plane and the surface (a); normal for the curvature radius definition of the given line (b); PQ—tangent.

c) The curvature radius of the normal intersection line, C_{norm}, at point M can be calculated by Euler's formula:

$$\frac{1}{R} = \frac{\cos^2 \alpha}{R_1} + \frac{\sin^2 \alpha}{R_2} \qquad (1.3.10)$$

where R_1, R_2 are radiuses of main curvatures (that will be defined below),

α is the angle between the intersection lines of the two main planes and the surface S at point M (these intersection lines are called main lines);

d) The main curvatures' radiuses are the maximum and minimum of R. They belong to the main planes. If a surface is given in a form $z = f(x, y)$, then it is possible to calculate the radiuses by finding the roots of the following equation:

$$(rt - s^2)\,R^2 + [2\,p\,q\,s - (1 + p^2)\,t - (1 + q^2)\,r]\,R + (1 + p^2 + q^2)^2 = 0 \quad (1.3.11)$$

where

$$p = \frac{\partial z}{\partial x}, \quad q = \frac{\partial z}{\partial y}, \quad r = \frac{\partial^2 z}{\partial x^2}, \quad t = \frac{\partial^2 z}{\partial y^2}, \quad s = \frac{\partial^2 z}{\partial x\, \partial y} \qquad (1.3.12a)$$

These coefficients, i.e., p, q, r, t, s, determine the surface curvature at some point, or the second quadratic shape of the surface:

$$G_{II} = L\,dx^2 + 2\,M\,dx\,dy + N\,dy^2 \qquad (1.3.12b)$$

where L, M and N are the coefficients of the second quadratic shape of the surface (section 1.3.4 below);

e) Intersection lines between the main planes and the surface are the main normal intersection lines;

f) Gaussian curvature of the surface at point M can be calculated as follows:

$$K \approx \frac{1}{R_1 R_2} \qquad (1.3.13)$$

where R_1 and R_2 are the main radiuses of curvature at this point (see Figures 1.2.2 and 1.2.3a);

g) If the signs of R_1 and R_2 at point M are equal, then the main intersection lines of the surface are concave or convex in the same direction, i.e., the surface is located on the same side of the tangent plane at this point. In this case point M is called elliptic point and the surface is called elliptic paraboloid. Gaussian curvature of an elliptic paraboloid is positive:

$$K > 0 \qquad (1.3.14)$$

If the signs of R_1 and R_2 at point M are different, then the main intersection lines of the surface are concave or convex in different directions. In this case the plain will not be tangent, but will intersect the surface at point M, which is called a hyperbolic point. The surface at this point is called hyperbolic paraboloid. Gaussian curvature of a hyperbolic paraboloid at point M is negative (see Figure 1.2.3b,c):

$$K < 0 \qquad (1.3.15)$$

If R_1 and R_2 at point M $\to \infty$, then one of the main lines is straight. In this case M is called parabolic point and the surface at this point is called parabolic surface (for example cylindrical surface). Such a surface has zero Gaussian curvature at point M:

$$K = 0 \qquad (1.3.16)$$

If a surface has the same Gaussian curvature at all points, i.e.,

$$K = \text{const} \qquad (1.3.17)$$

it is called surface with constant Gaussian curvature (a simplest example of such surface is a sphere);

h) The average surface curvature at point M is calculated as follows:

$$K_m = \frac{1}{2}\left(\frac{1}{R_1} + \frac{1}{R_2} \right) \qquad (1.3.18)$$

i) If a surface has zero average curvature at all points, it is called a minimal surface

$$K_m = 0 \ (R_1 = -R_2) \qquad\qquad (1.3.19a)$$

j) The shortest line between two points that belong to a surface is called a geodetic line.

If a spatial surface is given in an evident form $z = f(x, y)$, then a differential equation of a geodetic line has the following form:

$$(1 + p^2 + q^2)(d^2y/dx^2) = p\,t\,(dy/dx)^3 + (2\,p\,s - q\,t)(dy/dx)^2 + (p\,r - 2\,q\,s)(dy/dx) - q\,r \qquad (1.3.19b)$$

1.3.4 Coefficients of the second quadratic shape of the surface

Following Eq. (1.3.13), Gaussian curvature of a surface at some point can be obtained, using the following exact formula:

$$K = \frac{L\,N - M^2}{E\,G - F^2} \qquad\qquad (1.3.20)$$

where E, G and F are coefficients of the first quadratic shape (see expression 1.3.4);

L, N and M are coefficients of the second quadratic shape of the surface.

If the surface equation is given in an evident form $z = f(x, y)$, then

$$L = \frac{r}{\sqrt{1 + p^2 + q^2}}; \quad M = \frac{s}{\sqrt{1 + p^2 + q^2}}; \quad N = \frac{t}{\sqrt{1 + p^2 + q^2}} \qquad (1.3.21)$$

Expressions for q, r, s and t are given in Eq. (1.3.12). Substituting expressions of the first and second quadratic shapes into Eq. (1.3.20) yields:

$$K = \frac{r\,t - s^2}{(1 + p^2 + q^2)^2} = \frac{k_x\,k_y - k_{xy}^2}{(1 + p^2 + q^2)^2} \qquad (1.3.22)$$

where k_x and k_y are the surface curvatures at the given point in directions of axes x and y, respectively (normal curvatures that were defined above);

k_{xy} is the torsion curvature of the surface:

$$k_x = r = \frac{\partial^2 z}{\partial x^2}; \quad k_y = t = \frac{\partial^2 z}{\partial y^2}; \quad k_{xy} = s = \frac{\partial^2 z}{\partial x\,\partial y} \qquad (1.3.23)$$

1.3.5 Shallow surface

A surface is called shallow if the angle between the horizontal plane of its basis and tangent plane at any point that belongs to the surface does not exceed 18° (i.e., one fifth of 90°). Maximum rise of a shallow translation shell on rectangular basis of the shell should not exceed one fifth of the shorter basis length (Figure 1.3.2a). For revolution shallow shells on circular basis the maximum rise of the shell should not exceed one fifth of the basis diameter (Figure 1.3.2b).

The difference between the length of a curve (arc) on shallow surface and that of its horizontal projection is rather small. Similarly, the difference between geometric relations on the surface and those on its planar basis is also small. Therefore it is possible to select an orthogonal coordinates system for shallow shells $x = \text{const}$, $y = \text{const}$ instead of a curved coordinates system. In this case the coefficients of the shell's second quadratic shape are different from the curvature (normal and torsion) that were given in Figure (1.3.2b):

$$L = k_x;\ N = k_y;\ M = k_{xy} \tag{1.3.24}$$

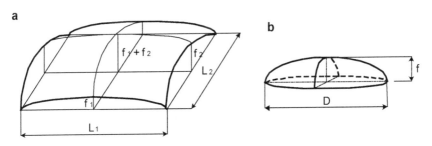

Figure 1.3.2. Shallow shells:

(a) translation surface [$L_2 < L_1$; $(f_1 + f_2) \le L_2/5$]; (b) revolution surface ($f \le D/5$).

1.4 Numerical examples in geometry of shells

1.4.1 Translation shell

Given a convex shallow translation rectangular shell in a form of an elliptic paraboloid with the following dimensions (Figure 1.2.3a):

$$L_1,\ L_2\ (a = L_1/2,\ b = L_2/2,\ L_1 \le 2\,L_2),\quad f_1 = 0.1\,L_1,\quad f_2 = 0.1\,L_2$$

The surface equation is:

$$z = 0.5\,(k_1\,x^2 + k_2\,y^2)$$

and the main curvatures are:

$$k_1 = 2f_1/a^2, \quad k_2 = 2f_2/b^2, \quad k_{12} = 0$$

For point M (x, y) that belongs to the surface calculate:

a) the first quadratic shape of the surface G_1 and its coefficients E, G and F.

b) the second quadratic shape of the surface G_{II} and its coefficients L, N and M.

c) the Gaussian curvature of the surface K.

Solution

a)

$$G_1 = E \, dx^2 + 2 \, F \, dx \, dy + G \, dy^2$$
$$E = 1 + p^2, \quad G = 1 + q^2, \quad F = p \, q$$
$$p = dz/dx = k_1 \, x, \quad q = dz/dy = k_2 \, y$$
$$E = 1 + k_1^2 \, x^2, \quad G = 1 + k_2^2 \, y^2, \quad F = k_1 \, k_2 \, x \, y$$
$$G_1 = (1 + k_1^2 \, x^2) \, dx^2 + 2 \, k_1 \, k_2 \, x \, y \, dx \, dy + (1 + k_2^2 \, y^2) \, dy^2$$

b)

$$G_{II} = L \, dx^2 + 2 \, M \, dx \, dy + N \, dy^2$$

Let's define

$$D = (1 + p^2 + q^2)^{0.5}$$

Then

$$L = r/D, \quad M = s/D, \quad N = t/D$$
$$r = d^2z/dx^2 = k_x, \quad s = d^2z/(dx \, dy) = k_{xy}, \quad t = d^2z/dy^2 = k_y$$

For a translation shell with constant curvatures x and y are the main directions, therefore

$$k_x = k_1, \quad k_2 = k_y, \quad k_{xy} = k_{12} = 0$$
$$G_{II} = (k_1 \, dx^2 + k_2 \, dy^2)/D$$

c) $K = (LN - M^2)/(EG - F^2) = [(rt - s^2)/D^2]/[(1 + k_y^2 x^2)(1 + k_x^2 y^2) - k_x^2 k_y^2 x^2 y^2]$

As the surface curvatures in the main directions x and y directions are constant, then

$$K = k_1 \times k_2 = 4 f_1 f_2/(a^2 \, b^2) > 0$$

1.4.2 Simple hyperboloid

Given a simple linear hyperboloid with rectangular basis dimensions, a and b, and the hyperboloid hog f (see Figure 1.2.3c). To be found: G_I, G_{II} and K.

Solution

The equation for the given shell's surface is:

$$z = f x y/(a b) = c x y, \quad c = f/(a b) = const$$

$$G_I = E \, dx^2 + 2 F \, dx \, dy + G \, dy^2$$

$$E = 1 + p^2, G = 1 + q^2, F = p q$$

$$p = dz/dx = c y, E = 1 + c^2 y^2$$

$$q = dz/dy = c x, G = 1 + c^2 x^2$$

$$F = c^2 x y$$

$$G_I = (1 + c^2 y^2) \, dx^2 + 2 c^2 x y \, dx \, dy + (1 + c^2 x^2) \, dy^2$$

$$G_{II} = L \, dx^2 + 2 M \, dx \, dy + N \, dy^2$$

$$L = r/D, M = s/D, N = t/D$$

$$D = (1 + p^2 + q^2)^{0.5} = (1 + c^2 y^2 + c^2 x^2)^{0.5}$$

$$r = d^2z/dx^2 = k_x = 0, \quad t = d^2z/dy^2 = k_y = 0, \quad s = d^2z/(dx \, dy) = k_{xy} = c = const$$

$$L = 0, \quad N = 0, \quad M = c/(1 + c^2 y^2 + c^2 x^2)^{0.5}$$

$$G_{II} = 2 M \, dx \, dy = 2 c/(1 + c^2 y^2 + c^2 x^2)^{0.5}$$

$$K = (L N - M^2)/(E G - F^2) = - [c^2/(1 + c^2 y^2 + c^2 x^2)]/[(1 + c^2 y^2) (1 + c^2 x^2) - c^4 x^2 y^2]$$

$$K < 0$$

1.4.3 Long cylindrical shell

Given a long cylindrical shell, its dimensions in plane are a and b, the rise of the shell is f (see Figure 5.1.1). To be found: G_I, G_{II} and K.

Solution

The equation of the shell's surface is:

$$z = f y^2/b^2 = c\, y^2, \ c = f/b^2$$

$$p = dz/dx = 0, \qquad q = dz/dy = 2\, c\, y$$

$$E = 1 + p^2 = 1, \qquad F = p\, q = 0, \qquad G = 1 + q^2 = 1 + 4\, c^2\, y^2$$

$$G_I = E\, dx^2 + 2\, F\, dx\, dy + G\, dy^2 = E\, dx^2 + G\, dy^2 = dx^2 + (1 + 4\, c^2\, y^2)\, dy^2$$

$$G_{II} = L\, dx^2 + 2\, M\, dx\, dy + N\, dy^2$$

$$r = d^2z/dx^2 = k_x = 0, \qquad t = d^2z/dy^2 = k_y = 2\, c, \qquad s = d^2z/(dx\, dy) = k_{xy} = 0$$

$$L = r/D = 0, \qquad M = s/D = 0, \qquad N = t/D = 2\, c/D = 2\, c/(1 + 4\, c^2\, y^2)^{0.5}$$

$$G_{II} = 2\, c\, dy^2/(1 + 4\, c^2\, y^2)^{0.5}$$

$$K = (L\, N - M^2)/(E\, G - F^2) = 0$$

1.4.4 Saddle-shaped shell

Given a translation rectangular shallow panel in a form of a saddle (hyperbolic paraboloid). The dimensions of the shell (Figure 6.1.1) are:

$$L_1,\ L_2\ (a = L_1/2, \qquad b = L_2/2, \qquad L_1 > 2\, L_2), \qquad f_1 = 0.1\, L_1, \qquad f_2 = 0.1\, L_2$$

The equation of the shell's surface is:

$$z = 0.5\, (k_1\, x^2 - k_2\, y^2)$$

and its main curvatures are:

$$k_1 = 2\, f_1/a^2, \qquad k_2 = 2\, f_2/b^2, \qquad k_{12} = 0$$

Point $M(x, y)$ belongs to the surface.

To be found for this point:

a) the first quadratic shape of the surface, G_I, and its coefficients E, G and F.

b) the second quadratic shape of the surface, G_{II}, and its coefficients L, M and N.

c) the Gaussian curvature of the surface K.

Solution

a)

$$G_1 = E\,dx^2 + 2\,F\,dx\,dy + G\,dy^2$$
$$E = 1 + p^2, \qquad G = 1 + q^2, \qquad F = p\,q$$
$$p = dz/dx = k_1\,x, \qquad q = dz/dy = -k_2\,y$$
$$E = 1 + k_1^2\,x^2, \qquad G = 1 + k_2^2\,y^2, \qquad F = -k_1\,k_2\,x\,y$$
$$G_1 = (1 + k_1^2\,x^2)\,dx^2 - 2\,k_1\,k_2\,x\,y\,dx\,dy + (1 + k_2^2\,y^2)\,dy^2$$

b)

$$G_{II} = L\,dx^2 + 2\,M\,dx\,dy + N\,dy^2$$

Let's define

$$D = (1 + p^2 + q^2)^{0.5}$$
$$L = r/D, \qquad M = s/D, \qquad N = -t/D$$
$$r = d^2z/dx^2 = k_x, \qquad s = d^2z/(dx\,dy) = k_{xy}, \qquad t = d^2z/dy^2 = -k_y$$

For a translation shell with constant curvatures, directions x and y are the main ones, therefore

$$k_x = k_1, \qquad k_2 = k_y, \qquad k_{xy} = k_{12} = 0$$
$$G_{II} = (k_1\,dx^2 - k_2\,dy^2)/D$$

c)

$$K = (L\,N - M^2)/(E\,G - F^2) = [(r\,t - s^2)/D^2]/[(1 + k_x^2\,x^2)\,(1 + k_y^2\,y^2) - k_x^2\,k_y^2\,x^2\,y^2]$$

As the curvatures of the given surface are constant in directions x and y, which are the main directions, then

$$K = k_1 \times k_2 = -4\,f_1\,f_2/(a^2\,b^2) < 0$$

1.4.5 Ellipsoid of revolution

Given a half ellipsoid of revolution by halves of its axes (Figure 1.4.1). The shell dimensions are:

$$a = 46.5\ \text{m}, \qquad b = 26\ \text{m}, \qquad c = 14\ \text{m}$$

a) For a point that belongs to the surface $z = f\,(x = 1\ \text{m}; y = 1\ \text{m})$ it is required to calculate the first quadratic shape of the surface, G_1, and its coefficients E, G and F.

b) What is the length of the segment, created by the intersection of the ellipsoid and a vertical plane $x = 10$ m, parallel to axis y, if the length of the segment's projection on the horizontal plane is 3 m and the segment center is located at point A ($x = 10$ m, $y = 10$ m)?

Solution

a)

$$x^2/a^2 + y^2/b^2 + z^2/c^2 = 1$$
$$z = [c/(a\,b)]\,(a^2\,b^2 - b^2\,x^2 - a^2\,y^2)^{0.5}$$
$$E = 1 + p^2, \qquad G = 1 + q^2, \qquad F = p\,q$$
$$p = dz/dx = -b\,c\,x/[a\,(a^2\,b^2 - b^2\,x^2 - a^2\,y^2)^{0.5}]$$
$$q = dz/dy = -a\,c\,xy/[b\,(a^2\,b^2 - b^2\,x^2 - a^2\,y^2)^{0.5}]$$
$$E = 1 + b^2\,c^2\,x^2/[a^2\,(a^2\,b^2 - b^2\,x^2 - a^2\,y^2)]$$
$$G = 1 + a^2\,c^2\,y^2/[b^2\,(a^2\,b^2 - b^2\,x^2 - a^2\,y^2)]$$
$$F = (c^2\,x\,y)/(a^2\,b^2 - b^2\,x^2 - a^2\,y^2)$$
$$x = y = 1 \text{ m}$$
$$E = 1 + b^2\,c^2/[a^2\,(a^2\,b^2 - b^2 - a^2)]$$
$$G = 1 + a^2\,c^2/[b^2\,(a^2\,b^2 - b^2 - a^2)]$$
$$F = c^2/(a^2\,b^2 - b^2 - a^2)$$

b)

$$L = (E\,\Delta x^2 + 2\,F\,\Delta x\,\Delta y + G\,\Delta y^2)^{0.5}$$

The segment length in y direction $\Delta y = 3$ m. Assuming that the breadth of the curve is zero, i.e., $\Delta x = 0$, yields

$$L = \Delta y\,G^{0.5}$$

For the given point A ($x = 10$ m; $y = 10$ m)

$$G = 1 + a^2\,c^2\,y^2/[b^2\,(a^2\,b^2 - b^2\,x^2 - a^2\,y^2)] = 1.053; \qquad G^{0.5} = 1.026$$

$$L = 3.078 \text{ m}$$

$$\Delta L = L - \Delta y = 0.078 \text{ m}$$

$$\Delta L/L = 0.0254 = 2.54\%$$

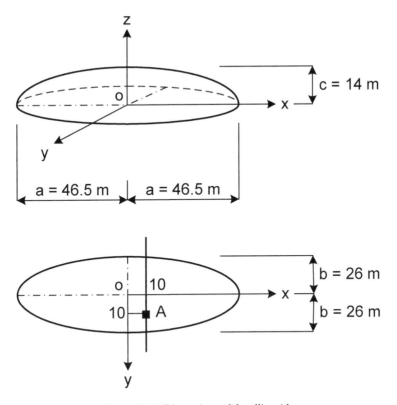

Figure 1.4.1. Dimensions of the ellipsoid.

1.4.6 Dome

Given a circular dome (see Figure 1.3.2b). The dome dimensions are:

$$r_0 = D/2 = 10 \text{ m}, \quad f = 3 \text{ m}, \quad R = 18.16 \text{ m}$$

Point A (3, 7) is located on the dome (Figure 1.4.2).

The equation of the dome's surface is:

$$z = (R^2 - x^2 - y^2)^{0.5}$$

a) Calculate the length of the intersection segment between the dome and the vertical plane $x = 3$ m, parallel to axis y, if the length of the segment projection on the horizontal plane is 2 m and the center of the segment is located at point A (3, 7).

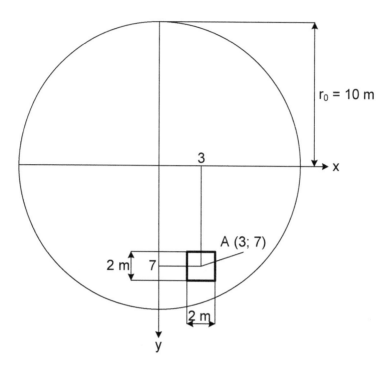

Figure 1.4.2. The given segment on the dome's surface (upper view).

b) Calculate a part of the dome's surface, if its horizontal projection is 2 × 2 m and its center is located at point A.

Solution

a)

$$p = dz/dx = x/(R^2 - x^2 - y^2)^{0.5}; \quad q = dz/dy = y/(R^2 - x^2 - y^2)^{0.5}$$

$$E = 1 + p^2 = (R^2 - y^2)/(R^2 - x^2 - y^2)$$

$$G = 1 + q^2 = (R^2 - x^2)/(R^2 - x^2 - y^2)$$

$$F = p\,q = x\,y/(R^2 - x^2 - y^2)$$

For $x = 3$ m, $y = 7$ m

$$E = 1.033, \quad G = 1.18, \quad F = 0.077$$

Assuming that the breadth of the curve is zero, i.e., $\Delta x = 0$, and knowing that $\Delta y = 2$ m, yields:

$$L \approx (E\,\Delta x^2 + 2\,F\,\Delta x\,\Delta y + G\,\Delta y^2)^{0.5} = \Delta y\,G^{0.5}$$

$$\Delta y = 2\text{ m}, \quad L = 2.17\text{ m}$$

The difference between the curved segment and its horizontal projection is:

$$\Delta\,(\%) = (2.17 - 2)\,100/2 = 8.6\%$$

b)

The area of the given segment is:

$$S = (E\,G - F^2)^{0.5}\,\Delta x\,\Delta y = (1.033\ 1.18 - 0.077^2)\,2\ 2 = 4.4\text{ m}^2$$

$$\Delta\,(\%) = (4.4 - 4)\,100/4 = 10\%$$

1.4.7 Cylindrical shell

Given a cylindrical shell (see Figure 5.1.1). The shell's dimensions are:

$$a = L_1/2 = 10\text{ m}, \quad b = L_2/2 = 8\text{ m}, \quad f = 3\text{ m}$$

To be found:

a) Area of the shell surface part, if its horizontal projection has dimensions of 2×2 m and a center, located at point A ($x = 2, y = 6$).

b) Length of a segment at the intersection between the shell and a plane $x = 2$ m, parallel to the y axis, if the length of this segment's projection on horizontal plane is 3 m and its center is located at point A.

Solution

a)

The radius of curvature for the given surface is:

$$R = (a^2 + f^2)/(2\,f) = 12.16\text{ m}$$

The surface equation is:

$$z = f\,y^2/b^2, \quad p = dz/dx = 0, \quad q = dz/dy = 2\,f\,y/b^2$$

$$E = 1 + p^2 = 1, \quad F = p\,q = 0, \quad G = 1 + q^2 = 1 + 4\,f^2\,y^2/b^4$$

For y = 6 m

$$G = 1.32$$

$$S = (E\,G - F^2)^{0.5}\,\Delta x\,\Delta y = G^{0.5}\,\Delta x\,\Delta y = 1.149\,\Delta x\,\Delta y$$

$$\Delta x = \Delta y = 2\ \text{m}, \qquad S = 4.6\ \text{m}^2$$

$$\Delta\,(\%) = (4.6 - 4)\,100/4 = 15\%$$

b)

The length of the segment $\Delta y = 2$ m at point A(2; 6) is:

$$L = G^{0.5}\,\Delta y = 1.149\ 2 = 2.3\ \text{m}$$

$$\Delta\,(\%) = (2.3 - 2)\,100/2 = 15\%$$

2

Structural Principles in Design of Spatial Concrete Structures

2.1 General

2.1.1 Definition of a spatial structure

When speaking about spatial structure geometry, it is common to suggest a surface that passes through the middle of any section of the shell (Figure 2.1.1a). Let's define specific features of a spatial structure. To obtain spatial behavior of a structure, subjected to external loads (the present book is focused on vertical loads), the structure should satisfy the following special requirements:

a) The structure should be continuous to allow the internal forces stream without disturbance from point to point in all directions, not just in that of external loads. The structure can have openings, in this case there are certain constructive requirements (see Section 2.1.3). Ribs or local increase in the structure's thickness do not break the structure's continuity, in spite that they lead to local stresses in connections between the structure and the ribs or/and in the zones with varied thicknesses, because the local stresses decrease very fast.

b) Thrust forces always appear in spatial structures. Their direction depends on the structure's curvature, or in other words, on the fact if the structure is concave or convex. If a structure is convex, the thrust forces act out the structure and lead to tension in elements that take them, i.e., ties or edge elements (Figure 2.1.1b); and vice versa, if a structure is concave, the thrust forces act in the inner direction ("negative" thrust) and lead to compression in elements that take them (Figure 2.1.1c). Such behavior corresponds to spatial structures with positive or zero Gaussian curvature (for example domes or cylindrical shells, respectively). In structures with negative Gaussian curvature (K < 0) the thrust forces yield both tension and compression (for example a saddle-shape shell—Figure 2.1.1d). The thrust problem is one of the main issues for spatial structures, because in order to take the forces appearing due to the thrust effect, the structure should have corresponding elements that are able to take these forces by tension or compression. It should be noted that thrust forces appear also in folders.

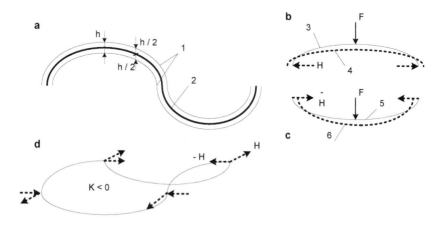

Figure 2.1.1. The problem of spatial structures definition:

a—the structure's thickness (1) and the middle surface (2); b—convex structure (3) and its thrust force (4); c—concave structure (5) and its "negative" thrust force (6); d—a saddle with "positive" and "negative" thrust forces; H—thrust force.

c) Spatial structures (except folders with flat plates) have some Gaussian curvature. The curvature can be single (like in cylindrical shells) or double (like in domes, or saddles (hypars)). In the first case the structure has zero Gaussian curvature (K = 0) and in the second one the curvature can be positive (K > 0) or negative (K < 0).

A spatial structure is usually a covering structure and therefore the live load is relatively small. The dominant load that acts on a spatial structure is its self-weight. If a spatial structure is a convex shell (K > 0) then from the viewpoint of statics most of its (about 80% of the area) is in compression. This is one of the main reasons to use reinforced concrete for construction of such structures. Therefore, it is logically to use thin-walled shells in order to reduce their self-weight. As the mass of spatial structures is rather small, their resistance to dynamic loadings, like earthquakes, is high. Following the shells' theory, spatial structure is thin-walled, if the ratio between its thickness, h, and the short span, l_{min} (for rectangular basis), or diameter (for circular basis) does not exceed 1/100:

$$\frac{h}{l_{min}} \leq \frac{1}{100}$$ (2.1.1)

Most existing shells are even thinner:

$$\frac{h}{l_{min}} \leq \frac{1}{200} ... \frac{1}{400}$$ (2.1.2)

For example, the average thickness of a positive Gaussian curvature shell, constructed by the first author (Iskhakov 1981) is 0.07 m and its basis is 24 × 24 m, i.e., $h/l \approx 1/340$. This shell will be discussed in details later.

At the same time, from a technological viewpoint, all RC spatial structures, h should be at least 5 cm.

2.1.2 Types of spatial structures and their structural elements

This book deals with the following types of spatial structures:

- translation shells with positive Gaussian curvature (K ≥ 0) on square or rectangular basis (Figure 2.1.2a);
- long cylindrical shells (K = 0) on rectangular basis (Figure 2.1.2b);
- domes–revolution shells (Figure 2.1.2c);
- hypar–shells with linear surface (negative Gaussian curvature, K < 0) on square or rectangular basis (Figure 2.1.2d);
- folders on rectangular basis (Figure 2.1.2e).

These types of shells are the most popular in the world. Two examples of such shells that collapsed in the recent years will be discussed in details in this book (Iskhakov and Ribakov 2014): a cylindrical roof shell of the new terminal at Charles de Gaulle Airport in Paris, France (long span length 650 m), and a dome above the Aqua-Park building in Moscow, Russia (126 m in perimeter).

Additionally, the biggest fiberglass dome in the world, the Millennium Dome in London, UK, with a diameter of 320 m (Liddel and Miller 1999) is described. One of the biggest timber geodetic domes with an RC ring that has a diameter of 106 m, constructed in Eilat, Israel (Abraham 2012), is analyzed. An ellipsoid RC shell in Chiasso, Switzerland (Muttoni et al. 2013) with dimensions in plane of 52 × 93 m is also discussed.

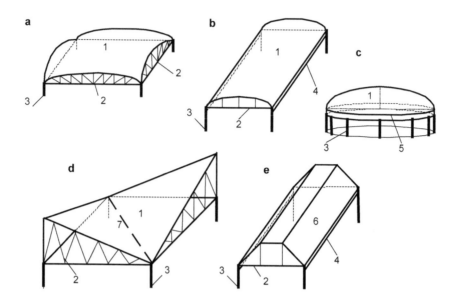

Figure 2.1.2. Types of shells and folders: a—translation shell with double curvature K > 0; b—cylindrical shell with single curvature K = 0; c—dome; d—hypar with double curvature K < 0; e—folder; 1—the shell (dome) itself; 2—diaphragm; 3—column; 4—edge element; 5—dome's ring; 6—the folder's plates; 7—tensile tie for taking thrust.

The main constructive elements of spatial structures are:

- the shell, dome or folder itself;
- a diaphragm to take the thrust forces from the shell or folder;
- an edge element that takes together with the shell or folder the bending moments in long direction;
- a ring that takes the thrust forces from a dome.

Types of diaphragms are shown in Figure 2.1.3. Diaphragms can be in a form of a truss (Figure 2.1.3a) or an arc with a tensile tie (Figure 2.1.3b). It is possible to transfer the thrust forces from a shell to a frame (Figure 2.1.3c), or even to the basis (Figure 2.1.3d). Sometimes end load carrying walls can be used for this purpose (Figure 2.1.3e). In this case the reaction from the shell is transferred along the entire length of the end load carrying walls (there are no thrust forces).

The edge elements usually have a rectangular section. Connection between the edge element and the shell depends on the boundary conditions of the structure. Schemes of typical connections are shown in Figure 2.1.4a. If it is required to provide water drainage from the roof then the connection has a form, shown in Figure 2.1.4b. Sometimes, if there are specific architectural requirements, the edge element is designed in a form of a horizontal plate

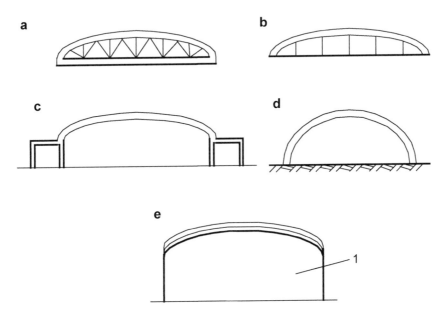

Figure 2.1.3. Types of diaphragms:

a—truss; b—arc with a tie; c—frame; d—basis; e—end load carrying wall (1).

(Figure 2.1.4c). In any case, the edge elements include longitudinal tensile reinforcement in the long direction of the shell. The edge elements decrease the vertical deflections of the structure in that direction.

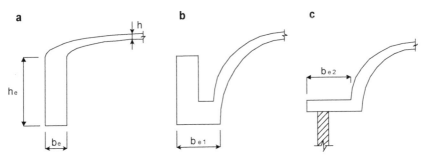

Notation: h—shell thickness; h_e and b_e—dimensions of the edge element's section; b_{e1} and b_{e2}—design dimensions.

Figure 2.1.4. Types of edge elements connection to the shell:

a—edge element, providing maximum inertia moment of the shell section; b—connection considering water drainage; c—shell supported on wall or beam.

2.1.3 Openings in shells

In most cases openings in shells have rectangular or circular shape. Connection of the opening to the shell depends on the dimensions of the first. If an opening is small relative to the shell dimensions, the local discontinuity has a minor effect on the stresses in the shell and can be neglected. If an opening is big, then the shell thickness is not enough to take the stresses concentrated along the opening. In this case, strengthening ribs are required along the perimeter of the opening. In both cases, additional reinforcement is provided along the perimeter of the opening, as shown in Figure 2.1.5.

According to engineering experience, an opening is small (Figure 2.1.5a), if

$$d \le 15\,h \qquad\qquad (2.1.3)$$

where d is the diameter of a circular opening or the longer dimension of a rectangular one;

h is the shell thickness.

And vise versa, an opening is big (Figure 2.1.5b), if

$$15\,h < d \le 30\,h \qquad\qquad (2.1.4)$$

To avoid additional concentrated stresses at the corners of rectangular openings they are rounded. The radius of the rounding curve should satisfy the following requirement:

$$r \ge 2\,h \qquad\qquad (2.1.5)$$

There also additional constructive requirements:

- additional reinforcement bar with a minimal diameter of 8 mm and minimal lap of 30∅ (∅ is the bar diameter) should be provided along the perimeter of a small opening;
- the ribs along the perimeter of a big opening should satisfy the following requirements:

$$h_1 \ge 3\,h; \quad b_1 \ge 2\,h; \quad 2\,b_1\,h_1 \ge d\,h \qquad\qquad (2.1.6)$$

- the section of the rib reinforcement

$$A_s \ge \rho\,d\,(h + h_1 - d_s) \qquad\qquad (2.1.7)$$

where ρ is the shell reinforcement ratio;

d_s is the thickness of the covering concrete layer,

h_1 is the bump of the rib (not including the shell thickness).

If the opening diameter is more than 30 h, it is necessary to perform exact static analysis of the shell in order to obtain the correct stresses distribution along the perimeter of the opening.

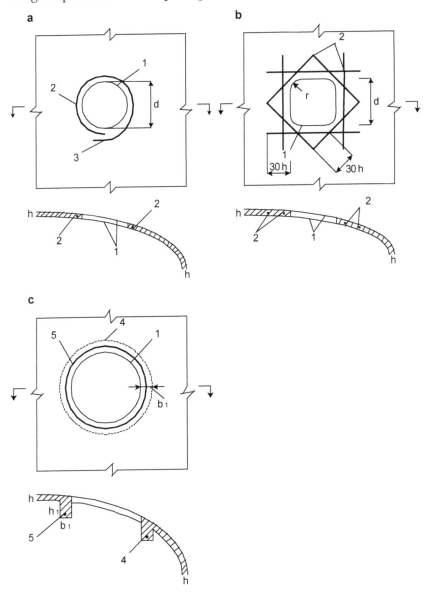

Figure 2.1.5. Openings in shells:

a and b—small openings; c—big opening; 1—perimeter of the opening 2—reinforcement bar along the perimeter; 3—the bar lap; 4—rib; 5—rib reinforcement.

2.2 Reinforcement schemes in shells and folders

2.2.1 Reinforcement types; using pre-stressed concrete; local bending moments

The reinforcement section of spatial structures can be obtained in the same two ways, like for regular RC ones:

- by calculation, but not less than the minimal reinforcement ratio;
- without calculations (according to constructive requirements) following modern design codes EN 1992.

The reinforcement types in spatial structures are also similar to ordinary RC elements:

- reinforcement nets in shell, dome or folder;
- carcass, consisting of longitudinal reinforcing bars and shear links in edge elements and diaphragms.

If the tensile stresses in edge elements of a spatial structure (due to axial or eccentric forces) are rather big and cause extensive cracking or big deflections, it is logical to consider the possibility for using pre-stressing. In diaphragms and edge elements pre-stressing is carried out using conventional methods. Pre-stressing of domes' foot rings, or at corners of shells (where the main tensile stresses usually appear) is more complicated because these elements are curved. An additional problem for post-pre-stressing of thin-walled shells is providing proper covering layer for the pre-stressing tendons and their bonding with concrete. Another important problem is transferring the pre-stressing forces' reaction to thin-walled shells (buckling problem) or to edge elements (as their out-of-plane stiffness is low).

Shells are thin-walled structures, whereas diaphragms have higher thickness. Therefore, the shell is to a certain degree jammed in the diaphragm. As a result, in addition to the membrane forces, bending moments appear along the perimeter of the shell. These moments decrease very fast as the distance from the perimeter increases. Therefore, it is called local moment. A reinforcing net (by calculation and not constructive) is required along the shell perimeter in order to take these moments.

2.2.2 Reinforcement schemes by calculation

Reinforcement schemes for translation shells with double curvature ($K > 0$) include the following reinforcement bars (Figure 2.2.1a):

- diagonal bars for the main tension at the corners;
- bars perpendicular to the diaphragms for the local bending moments;
- tensile bars in the diaphragms for the thrust forces.

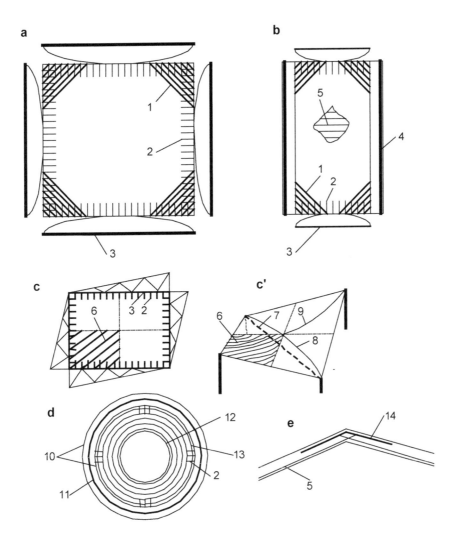

Note: 8, 9 and 12 are invisible lines, but not reinforcement.

Figure 2.2.1. Reinforcement of shells by calculation:

a—translation shell; b—cylindrical shell; c and c'—hypar; d—dome; e—folder; 1—diagonal bars; 2—reinforcement for local bending; 3—tensile bar in diaphragm; 4—longitudinal reinforcement of the edge element; 5—reinforcement positive for transverse moment; 6—diagonal reinforcement; 7—diagonal tensile bar; 8—diagonal arc (concave); 9—diagonal tendon (concave); 10—foot ring; 11—reinforcement for taking the thrust forces; 12—transferring ring; 13—reinforcement in dome rings under tension; 14—reinforcement for taking negative transverse moment in the top region of a folder.

Reinforcement schemes for long cylindrical shells with single curvature (K = 0) include the following reinforcement bars (Figure 2.2.1b):

- diagonal bars for the main tension at the corners;
- bars perpendicular to the diaphragms for the local bending moments;
- tensile bars in the diaphragms for the thrust forces;
- tensile bars in the bending edge elements;
- bars for the transverse moments in the shell (perpendicular to the edge element).

Reinforcement schemes for hypars with double curvature (K < 0) include the following reinforcement bars (Figure 2.2.1c and c'):

- bars for the main tension in the shell's concave diagonal direction;
- bars perpendicular to the diaphragms for local bending moments;
- an upper chord in diaphragms for the thrust forces;
- diagonal tensile bar.

Reinforcement schemes for revolution domes include the following reinforcement bars (Figure 2.2.1d):

- tension bars in the foot ring against the thrust forces;
- tensile reinforcement in the bottom rings of the dome up to the transferring ring (the dome's rings above the transferring one are in compression);
- reinforcement for taking the local bending moment along the dome perimeter.

Reinforcement schemes for long folders are similar to those of long cylindrical shells (Figure 2.2.1e):

- rods perpendicular to the diaphragms for the local bending moments;
- tensile bars in the diaphragms for the thrust forces;
- longitudinal bars in the edge elements;
- bars for taking the positive transverse moments in the folder;
- bars for taking the negative transverse moments in the folder.

2.2.3 Constructive reinforcement in shells, domes and folders

Constructive reinforcement in shells and folders provides minimal reinforcement of concrete. It is also used for connecting the main reinforcement bars (that are obtained by calculation) in reinforcement nets.

Translation shells:

- a reinforcement net at the entire shell surface according to minimal reinforcement requirements (one or two layers);

- transverse reinforcement bars, perpendicular to diagonal ones (pos. 1 in Figure 2.2.1a) at the shell's corners to connect the main reinforcement in a net;
- transverse reinforcement bars, perpendicular to the main ones, in order to connect the reinforcement for local bending moments in a net (pos. 2 in the figure).

Long cylindrical shells:

- transverse reinforcement bars, perpendicular to the diagonal ones at the shell's corners (pos. 1 in Figure 2.2.1b), to connect them in a net;
- transverse reinforcement bars, perpendicular to the main ones for local bending moments (pos. 2 in the figure);
- transverse reinforcement bars, perpendicular to the main ones, for the transverse moments (pos. 5 in the figure).

Hypars (linear surfaces):

- bars, perpendicular to the diagonal ones, at the entire shell (pos. 6 in Figure 2.2.1c). In practice, instead of a diagonal net, an orthogonal one is used (see section "Hypars");
- transverse reinforcement bars, perpendicular to the main ones for local bending moments (pos. 2 in the figure).

Domes:

- a reinforcement net in the dome surface between its top and the transferring ring (pos. 12 in Figure 2.2.1d), according to the minimal reinforcement requirements for compressed elements;
- bars in meridian direction perpendicular to the calculated parallel ones (pos. 13 in the figure), in order to connect them into a net;
- transverse reinforcement bars, perpendicular to the main ones for local bending moments (pos. 2 in the figure).

Long folders:

- transverse reinforcement bars, perpendicular to the main ones for local bending moments;
- transverse reinforcement bars, perpendicular to the main ones for transverse moments (pos. 5 in Figure 2.2.1e);
- same for negative transverse moments (pos. 14 in the figure).

3

Elements of Elastic Shells' Theory

3.1 Internal forces and deformations of thin-walled shells

3.1.1 Two groups of internal forces

The following types of internal forces appear in spatial structures under external loadings (Figure 3.1.1a, b):

a) normal forces N_x and N_y, acting in the plane of the element and causing its compression or tension;

b) horizontal shear forces N_{xy} that also act in the plane of the element;

c) bending moments M_v and M_y that act perpendicular to the element's plane;

d) vertical shear forces V_x and V_y that also act perpendicular to the element's plane;

e) torsion moments M_{xy} and M_{yx}.

It is possible to divide the above-mentioned five types of internal forces into two main groups—forces, acting in the plane of the elements, and forces that are perpendicular to the element's plane. The first group is called membrane forces (N_x, N_y and N_{xy}) and the second one, bending forces (V_x, V_y, M_x, M_y and M_{xy}).

The shells theory is based on the theory of elasticity, i.e., it is possible to apply the superposition principle for internal forces in spatial structures. Therefore, it is possible to represent the internal forces in an element of a spatial structure (Figure 3.1.1b) as a sum of the two above-mentioned groups of forces (Figure 3.1.1c, d). Such representation is logical first of all for thin-walled shells, because their bending resistance is very low. Actually, such shells are spatial membranes that take the external loadings just by internal membrane forces (N_x, N_y and N_{xy}).

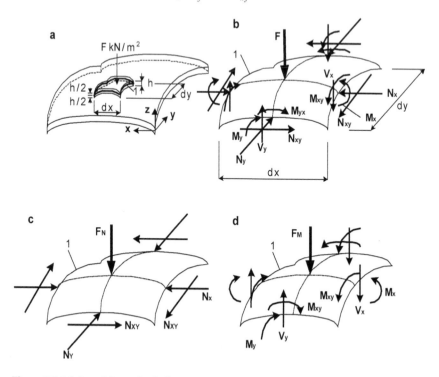

Figure 3.1.1. Internal forces in shells:

a—general view; b—general scheme of internal forces in a spatial differential element; c— membrane forces group; d—bending forces group; 1—middle surface of the shell.

As the in-plane stiffness of thin-walled membranes is much higher than the out-of-plane one, the following equations are true:

$$F = F_N + F_M;\ F_M \approx 0;\ F_N \approx F \tag{3.1.1}$$

where F is the external uniformly distributed load, kN/m²;

F_N is the part of the load, taken by the membrane forces group;

F_M is the part of the load, taken by the bending forces group (perpendicular to the membrane surface).

3.1.2 Local bending

As it was mentioned above, thin-walled shells have low stiffness at bending, relative to the membrane one. Therefore, the membrane forces are dominant over the main part of the shell (see Figure 3.1.1c) and the bending forces are very small and can be neglected. Hence, such shells have almost uniform deflections, except the regions close to the supports (edge elements), where the deflections decrease very rapidly up to zero and even change their sign (Figure 3.1.2a).

The edge elements are relatively thick and can be considered as semi-rigid supports for thin-walled shells (Figure 3.1.2b). As a result, negative bending moments usually appear in this region (local bending moments). According to available experimental data [1, 7], these moments rapidly decrease, and this region is rather narrow (up to 10% of the shell span—Figure 3.1.2c).

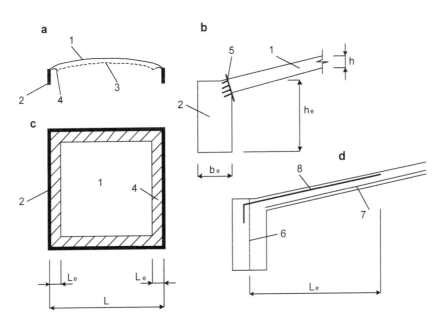

Notation: h—shell thickness; b_e, h_e—rectangular edge section dimensions; L—shell span; L_e—length of the local bending moment region.

Figure 3.1.2. Local bending moments in shells:

a—shell's section; b—the edge element as a semi-rigid support for shell; c—upper view on the local bending region; d—reinforcement nets in that region; 1—shell; 2—edge element; 3—almost uniform shell deflection; 4—local bending moment region along the perimeter of the shell; 5—semi-rigid support; 6—shell axis; 7—reinforcement net for membrane forces; 8—reinforcement net for local bending moments.

From another side, indeed because of rapid decrease in bending, it is possible to separate the influence of bending and membrane forces. Therefore, instead of considering a complex general internal forces' scheme, it is possible to calculate separately the bending and membrane forces and to sum them (see Figure 3.1.1d). For example, if reinforcement net 7 (Figure 3.1.2d) is required according to calculations to take the membrane forces, this net reaches the edge element that is a semi-rigid support for the shell. Reinforcement net 8, required according to calculations to take the negative local bending moment, is used additionally to net 7 in the same section.

To summarize, two separate calculations are carried out for thin-walled shells:

- calculating of the whole shell to membrane forces;
- calculating the region along the edge element to local bending moments.

The results obtained along the perimeter near the edge element for membrane forces and for the local bending elements are summed. This approach simplifies the calculations for RC shells as spatial structures and additionally (as it was demonstrated by many experimental investigations) it is rather accurate and recommended for practical applications (Iskhakov and Khaidukov 1996).

The width of the local bending moment region should satisfy the following requirement:

$$L_e \leq 0.1 \, L \tag{3.1.2}$$

3.1.3 Normal and bending deformations in translation shallow shells

Let's assume that $z_0 = f_0(x, y)$ and $z_d = f_d(x, y)$ are the equations of a shell middle surface before (at zero deformations stage) and after it is loaded by a uniformly distributed load (when some deformations appear), as shown in Figure 3.1.3. Let's define the first Gaussian square shape of the surface for the above-mentioned cases as G_{0I} and G_{dI}. Following Eq. (1.3.1),

$$G_I = ds^2 = E \, dx^2 + 2 \, F \, dx \, dy + G \, dy^2 \tag{3.1.3}$$

where E, F and G are the first shape coefficients.

The normal (membrane) deformation of the given surface is

$$\varepsilon = \frac{G_{0I} - G_{dI}}{G_{0I}} = 1 - \frac{G_{dI}}{G_{0I}} \tag{3.1.4}$$

Defining the second Gaussian square shape of the surface for the above-mentioned cases as G_{0II} and G_{dII} and using Eq. (1.3.12b),

$$G_{II} = L\ dx^2 + 2\ M\ dx\ dy + N\ dy^2 \tag{3.1.5}$$

where L, M and N are the second shape coefficients.

It is possible to calculate the bending deformations (relative difference in the Gaussian curvature of the surface) as follows:

$$\Delta K = \frac{G_{0II} - G_{dII}}{G_{0I}} \tag{3.1.6}$$

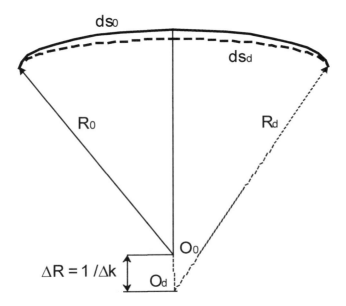

Notation: O_o and O_d—centers of curvature of the shell sections before (ds_0) and after (ds_d) the load application; R_o and R_d—corresponding radiuses of curvature (corresponding curvatures are k_0 and k_d).

Figure 3.1.3. The shell section before (solid line) and after (dashed line) bending deformations.

Example

Given a middle surface of a square translation shallow shell with constant positive Gaussian curvature in both main directions:

$$z = 0.5\ k\ (x^2 + y^2) \tag{3.1.7}$$

The first and the second Gaussian square shapes of the surface, following Eqs. (3.1.3) and (3.1.5), are:

$$G_I = ds^2 = E\, dx^2 + G\, dy^2 + F\, dx\, dy \tag{3.1.8}$$

$$E = 1 + p^2; \quad G = 1 + q^2; \quad F = p\,q; \quad p = \frac{\partial z}{\partial x} = \frac{dz}{dx} = k\,x; \quad q = k\,y;$$

$$E = 1 + k^2 x^2; \quad G = 1 + k^2 y^2; \quad F = p\,q = (dz/dx)\,(dz/dy) = k^2\,x\,y$$

$$G_I = ds^2 = (1 + k^2 x^2)\, dx^2 + (1 + k^2 y^2)\, dy^2 + k^2\, x\, y\, dx\, dy \tag{3.1.9}$$

$$L\frac{r}{D}; \quad N\frac{t}{D}; \quad M = \frac{s}{D}; \quad D = \sqrt{1 + p^2 + q^2}\,;$$

$$r = \frac{\partial^2 z}{\partial x^2} = \frac{d^2 z}{dx^2} = k; \quad t = \frac{d^2 z}{dy^2} = k; \quad s = d^2 z/(dx\, dy) = 0;$$

$$D = \sqrt{1 + k^2\, x^2 + k^2\, y^2}$$

$$G_{II} = \frac{k\,(dx^2 + dy^2)}{\sqrt{1 + k^2\, x^2 + k^2\, y^2}} \tag{3.1.10}$$

According to Eqs. (3.1.4) and (3.1.9),

$$\varepsilon\,(x, y) = 1 - \frac{(1 + k_d^2\, x^2)\, dx^2 + (1 + k_d^2\, y^2)\, dy^2}{(1 + k_0^2\, x^2)\, dx^2 + (1 + k_0^2\, y^2)\, dy^2} \tag{3.1.11}$$

As translation shells have no torsion curvature and assuming that Poisson deformations are negligible, the internal forces in x and y directions are independent each of the other. Therefore, a function of two variables $\varepsilon\,(x, y)$ can be represented by two single-variable functions $\varepsilon\,(x)$ and $\varepsilon\,(y)$:

$$\varepsilon(x) = \frac{x^2\,(k_0^2 - k_d^2)}{1 + k_0^2\, x^2}; \quad \varepsilon(y) = \frac{y^2\,(k_0^2 - k_d^2)}{1 + k_0^2\, y^2} \tag{3.1.12}$$

According to Eqs. (3.1.6) and (3.1.10),

$$\Delta K(x, y) = \frac{\dfrac{k_0\,(dx^2 + dy^2)}{\sqrt{1 + k_0^2\, x^2 + k_0^2\, y^2}} - \dfrac{k_d\,(dx^2 + dy^2)}{\sqrt{1 + k_d^2\, x^2 + k_d^2\, y^2}}}{\left(1 + k_0^2\, x^2\right) dx^2 + \left(1 + k_0^2\, y^2\right) dy^2} \tag{3.1.13}$$

For a translation shell:

$$\Delta k(x) = \frac{\dfrac{k_0\, dx^2}{\sqrt{1+k_0^2\, x^2}} - \dfrac{k_d\, dx^2}{\sqrt{1+k_d^2\, x^2}}}{\left(1+k_0^2\, x^2\right) dx^2} =$$

$$= \frac{k_0}{(1+k_0^2\, x^2)^{3/2}} - \frac{k_d}{(1+k_0^2\, x^2)(1+k_d^2\, x^2)^{1/2}}$$ (3.1.14)

$$\Delta k(y) = \frac{k_0}{(1+k_0^2\, y^2)^{3/2}} - \frac{k_d}{(1+k_0^2\, y^2)(1+k_d^2\, y^2)^{1/2}}$$ (3.1.15)

where k_0 and k_d are the surface main curvatures before and after deformations.

3.2 Equilibrium equations of the shell element

3.2.1 Components of the deflections' vector

Let's assume that the deflections vector at some point M, belonging to a middle surface of the shell, is MM′ (Figure 3.2.1). The components of this vector in x, y and z directions are u, v and w, respectively.

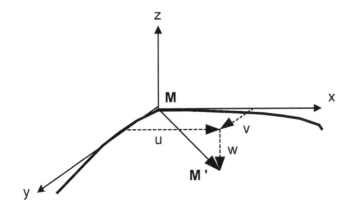

Figure 3.2.1. Components of vector MM′.

It is proved in the theory of elastic shallow shells that there are the following relations between deflections and deformations:

$$\varepsilon_x = \frac{\partial u}{\partial x} - \Delta k_x\, w$$ (3.2.1)

$$\varepsilon_y = \frac{\partial v}{\partial y} - \Delta k_y \, w \qquad (3.2.2)$$

$$\varepsilon_{xy} = \frac{\partial u}{\partial y} + \frac{\partial v}{\partial x} - \Delta k_{xy} \, w \qquad (3.2.3)$$

$$\Delta k_x = -\frac{\partial^2 w}{\partial x^2} \qquad (3.2.4)$$

$$\Delta k_y = -\frac{\partial^2 w}{\partial y^2} \qquad (3.2.5)$$

$$\Delta k_{xy} = \frac{\partial^2 w}{\partial x \, \partial y} \qquad (3.2.6)$$

There are also relations between deformations and internal forces:

$$N_x = \frac{E\,h}{(1-v^2)}(\varepsilon_x + v\,\varepsilon_y) \qquad (3.2.7)$$

$$N_y = \frac{E\,h}{(1-v^2)}(\varepsilon_y + v\,\varepsilon_x) \qquad (3.2.8)$$

$$N_{xy} = \frac{E\,h}{2\,(1+v)}\,\varepsilon_{xy} \qquad (3.2.9)$$

where v is the Poisson's coefficient.

As the Poisson's coefficient for reinforced concrete is rather small ($v = 0.16 \ldots 0.2$), then $v^2 \ll 1$. Hence it is possible to simplify the above-mentioned expressions:

$$N_x = E_c\,h\,\varepsilon_{x'}, \quad N_y = E_c\,h\,\varepsilon_{y'}, \quad N_{xy} \approx 0.5\,E_c\,h\,\varepsilon_{xy} \qquad (3.2.10)$$

where E_c is the concrete elasticity modulus (according to the concrete class).

Correspondingly, the simplified relations for the bending forces group are:

$$M_x = -E_c\,I\,\Delta k_{x'}, \quad M_y = -E_c\,I\,\Delta k_{y'}, \quad M_{xy} = -E_c\,I\,\Delta k_{xy'}, \quad I = 1\,h^3/12 \qquad (3.2.11)$$

where I is the inertia moment of the shell section (for a 1 m width strip).

3.2.2 Equilibrium equations of a thin-walled shallow shell element

A system of differential equations for an element of a thin-walled shallow shell includes three equations. The number of equations is equal to that of unknowns—normal membrane forces N_x, N_y and membrane shear force N_{xy}, i.e., there is a close theoretical solution for this system. The differential equations are:

$$\frac{\partial N_x}{\partial x} + \frac{\partial N_{xy}}{\partial y} = 0$$

$$\frac{\partial N_{xy}}{\partial x} + \frac{\partial N_y}{\partial y} = 0$$

$$k_x N_x + k_y N_y + 2\,k_{xy} N_{xy} = -q \qquad\qquad (3.2.12)$$

where q is the external symmetric normal load.

For translation shells the torsion curvature equals to zero, i.e., $k_{xy} = 0$. Therefore, the expressions can be simplified as follows:

$$\frac{\partial N_x}{\partial x} + \frac{\partial N_{xy}}{\partial y} = 0$$

$$\frac{\partial N_{xy}}{\partial x} + \frac{\partial N_y}{\partial y} = 0$$

$$k_x N_x + k_y N_y = -q \qquad\qquad (3.2.13)$$

The solution of the system, i.e., finding the internal forces N_x, N_y and N_{xy}, depends on the shell's type and its boundary conditions. These issues will be discussed in the following sections.

3.2.3 Boundary conditions

Generally, shells are supported on planar edge elements that have length, appropriate to the shell span. Such elements have low out-of-plane stiffness. Moreover, in most cases the edge elements are made of steel and not of reinforced concrete, as they are subjected to tension due to the horizontal shear forces, transferred by the shell (Figure 3.2.2). In this case, there is no need of pre-stressing concrete and the structure has lower self-weight, which is especially important in seismic regions.

Nxy

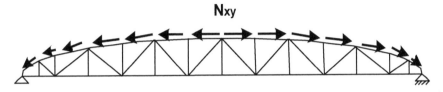

Figure 3.2.2. An edge element (diaphragm), subjected to tension due to the shell horizontal shear forces.

The in-plane stiffness of edge elements is rather high, therefore for simplicity reasons its in-plane deformations can be neglected. Therefore it is usually assumed in static calculations that edge elements are ideal from the viewpoint of deformations: insufficiently stiff in their plane (vertical direction) and insufficiently flexible in the out-of-the-plane (horizontal direction), as shown in Figure 3.2.3.

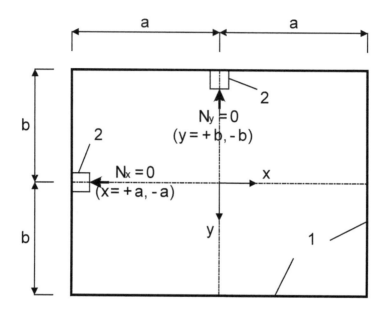

Figure 3.2.3. Idealized boundary conditions for a non-continuous shell (upper view):

1—edge element; 2—shell element near the diaphragms.

The following idealized boundary conditions are valid along the perimeter of the shell:

$$x = \pm a, \quad N_x = 0, \quad w = 0 \tag{3.2.14}$$

$$y = \pm b, \quad N_y = 0, \quad w = 0 \tag{3.2.15}$$

4

Convex Translation Shells

4.1 Reinforced concrete shells with steel trusses

4.1.1 General

The use of spatial convex translation shells in construction increases from year to year. The main reasons for this is that there is a need to cover longer spans of buildings, and such systems have become attractive architectural objects that dominate in city centers.

Engineers prefer to use common construction material, like concrete in spatial structures. It is clear that compression stresses are preferable in RC shells under external loads. Therefore, shells with positive Gaussian curvature are more suitable, as most of their area (up to 80%) is in compression (see Figure 4.2.5). At the same time the edge elements (diaphragms) in such shells are in tension (as they should take the thrust forces), hence the most appropriate material for these elements is steel. Combination of steel and reinforced concrete is the main idea for such shells. It allows construction of spatial structures with rather low self-weight that are effective for ordinary conditions and for regions with high seismicity.

In this chapter, examples of RC shells with steel trusses are discussed. Dimensions of these shells are 18 × 18 m, 24 × 24 m and 30 × 30 m. For spans shorter than 18 m planar systems can be used, and for spans longer than 30 m there is a problem of cracking due to the main tensile (diagonal) stresses at the corners that requires concrete pre-stressing, which is rather complicated for thin-walled shells. Therefore, the range of optimal spans for using the above-mentioned non-pre-stressed RC shells is between 18 and 30 m. Other shell dimensions within these limits (21 × 21 m, 27 × 27 m, 24 × 30 m, 18 × 24 m, etc.) are also possible.

4.1.2 The shell structure

The shells represented in the examples below are thin-walled, shallow and translation ones with positive Gaussian curvature (Figure 4.1.1a). The creating line rise is equal to that of the tracks lines ($f_1 = f_2 = f$) and for square shallow shells it is one tenth of their span. As a result, the rise at the center of the shell is equal to $2f$ (one fifth of the span).

Let's allocate the origin of the coordinates system XOZ at the center of the shell, so that the Z-axis would be upwards. In this case, the equation, describing the shell surface, is:

$$z = 0.5\,k\,(x^2 + y^2); \quad k = 2\,f/(L/2)^2; \quad f = L/10 \tag{4.1.1}$$

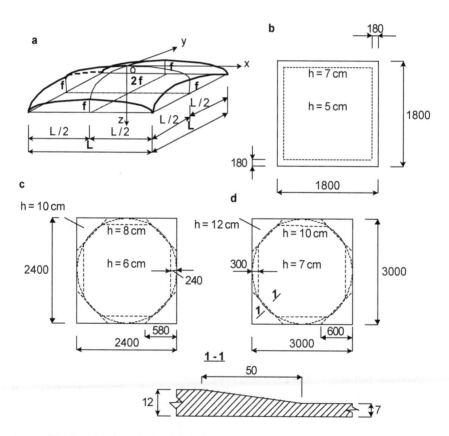

Figure 4.1.1. Details of translation RC shells:

a—shells geometry; b—the 18 × 18 m shell; c—the 24 × 24 m shell; d—the 30 × 30 m shell.

Actually, the RC shell is a monolithic curved convex membrane with minimal reinforcement, made of concrete class C30 or higher with cone slump of 6–8 cm. For shells without external water or/and thermal isolation it is recommended to consider non-cracked concrete with water proof at least B8. Concrete should be produced using Portland cement class 400–600, sand fraction up to 5 mm and aggregate fractions up to 20 mm.

The shell thin-walled and distribution of the thickness over the shell surface is presented in Figure 4.1.1b, c, d. The average thickness is 6.0, 7.4 and 8.8 cm for shells with 18 m, 24 m and 30 m spans, respectively (1/300–1/350 of the span). The covering concrete layer depth at the corners and along the perimeter is minimum 20 mm, whereas at the central part of the shell (subjected to compression) it is minimum 15 mm (but can be increased according to the environmental conditions).

The constructive reinforcement in the central part, near the upper and lower shell surface (in seismic regions) is a welded wire net Ø5 @ 20. At the corners and along the perimeter there are also reinforcement nets (by calculation), according to bending moments and main tensile stresses.

The edge elements (diaphragms) are steel trusses (Figure 4.1.2a) with UNP 30 at the upper belt and two corner profiles in all other elements. The flanges of the U—profiles are in the upward direction, in order to enable to cast the monolithic concrete connecting the diaphragm and the shell (Figure 4.1.2b). The horizontal shear forces, N_{xy}, are transferred from the shell to the upper belt of the diaphragm by welded corner profiles (Figure 4.1.2c). The distance between the corner profiles is calculated, but it should be up to 50 cm at the supports regions (0.25 L from each side of the support) and 100 cm in the remaining ones.

To prevent de-bonding of the shell concrete from the diaphragm steel upper belt, additional Ø16 links are welded between the corner profiles (Figure 4.1.2d). The upper view of this connection is shown in Figure 4.1.2e. As it is technically complicated to install an ordinary reinforcement net for taking the negative moments along the shell perimeter, reinforcing bars (pos. 8 in Figure 4.1.2f) are used instead, which are unified by a separate longitudinal bar (pos. 9 in the figure). The section and length of these bars (pos. 8) are calculated according to the local moments along the perimeter of the shell, but not less than at the anchorage length, l_{an}.

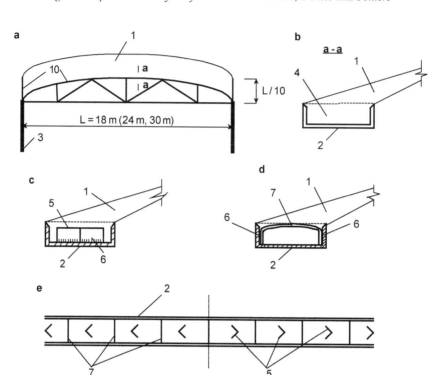

Figure 4.1.2. Detailed design of the connection between the shell concrete and the upper belt of the steel truss:

a—RC shell (1) and steel truss (10) on column supports (3); b—UNP 30 profile (2) as casting formwork (4); c—corner profiles (5) for taking the horizontal shear forces; welds (6) to the U profile, d—links (7) against concrete de-bonding; e—upper view of the connection; f—reinforcing bars (8) for taking the negative local moments, and (9)—longitudinal Ø12 bar for welding bars 8.

4.1.3 Supports of trusses

All contour trusses, used as shell edge elements (diaphragms), independent on the spans (from 18 to 30 m), have a 600 mm structural height at the supports for convenient construction of the connection between the diaphragms and columns (Figure 4.1.3a). The support itself is a vertical 12 mm thickness steel plate, strengthened by at least 14 mm ribs. The two corner profiles of the bottom belt are welded to the above-mentioned plate from the outer side of the column (Figure 4.1.3b).

A 20 mm horizontal steel plate with special orifices for further connection to columns is welded to the support before placing the truss. After putting the truss, each plate is bolted to column. There is a common bolt for two trusses at the shell corners. In other words, there are three bolts at each shell corner (Figure 4.1.3a). Additional 75 × 6 mm corner profiles are welded from both sides to the adjacent trusses vertical plates.

The shell is supported at the corners by RC columns with section (or head) dimensions of 50 × 50 cm. From the viewpoint of statics, the columns

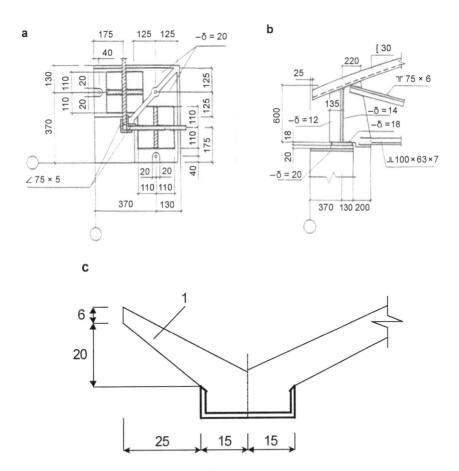

Figure 4.1.3. Shell supporting system and water drainage:

a—connection between the bottom belts of adjacent trusses; b—front view; c—water drainage from the shell by visor (1); δ-steel plate thickness.

act as hinged-fixed supports. It should be noted that the center of the trusses' supporting system is the connection point between the upper belt and the first diagonal rod (as the first element of the bottom belt is free of stresses). Therefore, the supporting region height of the truss is a part of the total height of the shell.

Water drainage from the shell at the end of a building is provided, using a cantilever visor that is casted monolithically with the shell (Figure 4.1.3c). It has a 25 cm length, 6 cm thickness (that increases uniformly in the shell's direction) and a 20 cm rise additionally to the thickness. It is necessary to install such visors from each side of a single-span shell. In a multi-span system water drainage is provided by downspout pipes.

4.1.4 Supporting of shells by row of columns

Shells, used as roofs (for low weight sport or cultural structures, storages, etc.), are supported at their corners by columns (as shown in Figure 4.1.2a). If there are walls or rows of columns, it is possible to use them as load bearing elements, instead of trusses, and to support the shell on them (Figure 4.1.4a). In such a case, just the upper belts of the edge trusses (diaphragms)

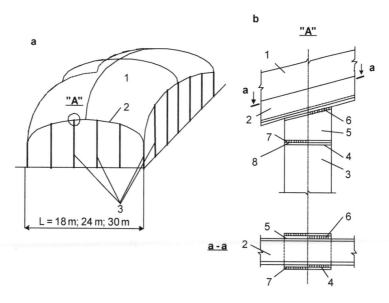

Figure 4.1.4. Supporting of a shell on row of columns:

a—general 3D view; b—connection detail "A"; 1—shell; 2—upper belt of the truss; 3—column; 4—column head; 5—wedge; 6—welding to the profile; 7—welding to the column head; 8—12 mm thickness steel plate anchorage in the column head.

are used (see Figure 4.1.3a, b). The intermediate columns carry the loads, transferred from the upper belt by wedge elements (Figure 4.1.4b) that act as hinges. These hinges are fixed supports. The external part of the wedge is welded to the upper belt profile and the internal one—to the column head.

4.1.5 Recommendations for shell construction

The order of operations for shell construction is:

- columns' placement;
- placing the edge steel trusses (diaphragms);
- constructing the formwork, according to the lower surface of the shell;
- placing the shell reinforcement;
- shell casting (measures against concrete suction to the formwork should be taken in order to prevent problems, related to removing the formwork after concrete hardening);
- concrete curing and hardening (at least up to 70% of the design strength);
- removing the formwork.

   The casting process should be continuous until the entire shell area is covered by concrete. Casting should start from the corners and perimeter to the shell center, in order to prevent the concrete mix creep. Appropriate measures (as a rule, water drenching) should be taken to prevent concrete cracking due to its shrinkage, because a thin concrete layer is distributed along a big shell area.

4.1.6 The design loads and materials' consumption

The vertical live load that acts on the shell was assumed to be 2.5 kN/m² (additionally to the self-weight). The shell was designed, considering seismic load, calculated according to peak ground acceleration of 0.3 g (where g is gravity acceleration). The horizontal forces are taken by the "shell–columns" system, acting as a spatial frame. It can be assumed that, as columns are flexible compared to the diaphragms, they take these forces alone as cantilevers (the connection between the diaphragms and the columns is hinged). As the shell span is long, the vertical component of the seismic load should also be considered.

   The concrete and cement consumptions are shown in Table 4.1.1. Steel consumption is presented in Table 4.1.2.

Table 4.1.1. Concrete and cement consumptions for construction of shells.

Dimensions of shells, m	Concrete class	Concrete thickness, cm	Concrete volume, m³	Cement consumption (class 400), t
18 × 18	C 30	6.0	19.71	7.88
24 × 24	C 30	7.4	43.03	17.21
30 × 30	C 30	8.8	79.29	31.72

Table 4.1.2. Steel consumption for construction of shells (kg).

Dimensions of shells, m	Reinforcing steel	Profiles' steel	Total	Per m²
18 × 18	2015.3	6629.2	8644.5	26.68
24 × 24	4654.6	11241.3	15805.9	27.44
30 × 30	6870.3	16775.2	23645.5	26.27

4.1.7 Example of a shell supported on row of columns

An example of a shell, supported on row of columns, is a building of the Museum of Modern Culture (MUMOK) in Vienna, Austria. The museum is located in a historical center of the city. The area was previously used as a stable for the Austrian imperator's horses in the Baroque period. Today it is a cultural and entertainment center (Figure 4.1.5).

Figure 4.1.5. The MUMOK building, Vienna, Austria (the covering shell is supported on rows of columns).

In the sixties of the previous century, part of the historical buildings in this area were adapted to a cultural museum. In 2001 a new building was designed by Ortner & Ortner architectural office (Germany). The building dimensions are 44 × 31 m, the height of the columns' rows from the ground up to the horizontal plane of the shell corners is 27.5 m, the maximum height of the structure is 35 m (Figure 4.1.6). The building is made of reinforced concrete and faced by basalt stone. The roof of the building is a rectangular translation thin-walled shell, supported by 28 columns, located along the perimeter (there are no interior columns in the building's space). The shell rise at the center is 7.5 m, the rise at the short and long edge sides are 3.15 m and 4.35 m, respectively. In other words, the shell is shallow in both long and short directions ($f_i/L_i = 1/10$, $i = 1; 2$).

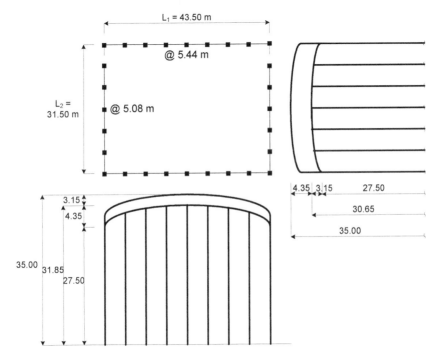

Figure 4.1.6. Columns' plane and dimensions of the shell, supported on rows of columns (cultural museum, Vienna).

4.1.8 Example of roofing shell with reinforced concrete arc diaphragms

Another RC shell was constructed in Tallinn, Estonia, in the sixties of the previous century (Figure 4.1.7). The shell is made of pre-cast 3 × 3 m RC elements. The structure is 40 × 40 m in plane, its rise from each side is 4 m. It is supported on pre-stressed concrete diaphragms that have a form of 40 m length arcs. The shell is a roof of a public building (market).

Figure 4.1.7. Roofing shell in Tallinn (Estonia) with arc diaphragms.

4.2 Calculation of internal membrane forces in shallow rectangular shells

4.2.1 Function of stresses

A system of differential equations for calculating the internal forces (3.2.12) is:

$$\frac{\partial N_x}{\partial x} + \frac{\partial N_{xy}}{\partial y} = 0$$

$$\frac{\partial N_{xy}}{\partial x} + \frac{\partial N_y}{\partial y} = 0$$

$$k_x N_x + k_y N_y + 2 k_{xy} N_{xy} = -q \qquad (4.2.1)$$

where q is a uniformly distributed load that acts on the shell area, kN/m².

This system can be solved, using a stresses function $\varphi\,(x, y)$ that is inter-related with the internal forces by the following equations:

$$N_x = \frac{\partial^2 \varphi}{\partial y^2}; \quad N_y = \frac{\partial^2 \varphi}{\partial x^2}; \quad N_{xy} = -\frac{\partial^2 \varphi}{\partial x\, \partial y} \qquad (4.2.2)$$

In this case, the first and second equations in the system become an identity and just the third equation remains valid and should be solved:

$$k_x N_x + k_y N_y + 2 k_{xy} N_{xy} = -q \qquad (4.2.3)$$

There is no exact solution of this equation for rectangular shells, there are just approximate methods. The stresses' function should satisfy the third above-mentioned equation and the boundary conditions of the shell.

Let's assume an approximate solution, for which the stresses' function is given in a form of algebraic polynomial:

$$\varphi(x, y) = \sum_i a_i \varphi_i(x, y) \qquad (4.2.4)$$

where a_i are constant parameters;

$\varphi_i(x, y)$ are the algebraic polynomial's elements;

i is the element's number.

In a case when the shell itself and the loads that act on the shell are symmetric, it is more convenient to assume the stresses' function is a product of two single-variable functions:

$$\varphi_i(x, y) = \varphi_i(x) \varphi_i(y) \qquad (4.2.5)$$

i.e., Eq. (4.2.3) will take the following form:

$$\varphi(x, y) = \sum_i a_i \varphi_i(x) \varphi_i(y) \qquad (4.2.6)$$

Generally, functions φ_i and parameters a_i should satisfy the boundary condition and the third equation of the system (4.2.1), respectively.

A known approximate method that provides exact solution at 3 or 4 selected shell points ($i = 3$, or $i = 4$) is a co-location method. When this method is applied, the stresses' function (4.2.6) and expressions for internal forces (4.2.2) are substituted into Eq. (4.2.3). As a result, i equations with i unknowns a_i are obtained.

4.2.2 Calculation of rectangular shells with infinitely stiff edge elements

A structural scheme of shells with infinitely stiff edge elements is shown in Figure 4.2.1. Let's analyze theoretically the edge element (that is usually a diaphragm in a form of truss) with infinitely high in-plane stiffness (vertical direction) and infinitely flexible out-of-plane stiffness (horizontal direction). In such a case, the boundary condition is (see Figure 4.2.1):

$$N_x \big|_{x = \pm a} = 0, \quad N_y \big|_{y = \pm b} = 0, \quad N_{xy} \big|_{x = \pm a, y = \pm b} \neq 0 \qquad (4.2.7)$$

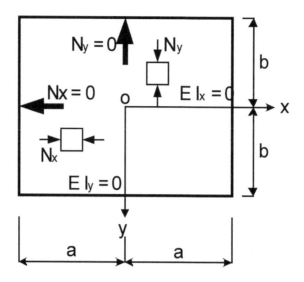

Figure 4.2.1. Edge elements (diaphragms) with infinite flexibility in the out-of-plane direction.

The stresses' function at the edges for symmetric external loading and system of coordinates should be an even function, i.e., it should be represented by an exponential algebraic polynomial. Therefore, just elements with even exponents should remain. The function should be selected so that its first term $\varphi_1(x, y)$ satisfies the solution of Eq. (4.2.3) in first approximation, and the other terms $\varphi_2(x, y)$, $\varphi_3(x, y)$, ..., $\varphi_i(x, y)$ improve the solution accuracy. For realizing these requirements for $i = 3$, it is possible to assume the stresses' function in the following form:

$$\varphi_i(x, y) = \varphi_i(x)\, \varphi_i(y) = a_1\, A_{x1}\, A_{y1} + a_2\, A_{x2}\, A_{y1} + a_3\, A_{x1}\, A_{y2} \qquad (4.2.8)$$

where

$$A_{x1} = x^4 - 6\, x^2\, a^2 + 5\, a^4$$

$$A_{y1} = y^4 - 6\, y^2\, b^2 + 5\, b^4$$

$$A_{x2} = x^8 - 2.444\, x^6\, a^2 + 1.444\, x^4\, a^4$$

$$A_{y2} = y^8 - 2.444\, y^6\, b^2 + 1.444\, y^4\, b^4$$

The internal membrane forces can be calculated according to Eq. (4.2.2):

$$N_x = 12\, a_1\, A_{x1}\, (y^2 - b^2) + 12\, a_2\, A_{x2}\, (y^2 - b^2) + 4\, a_3\, A_{x1}\, B_y \qquad (4.2.9)$$

where

$$B_y = 14\, y^6 - 18.333\, y^4\, b^2 + 4.333\, y^2\, b^4$$

$$N_y = 12\, a_1\, A_{y1}\, (x^2 - a^2) + 12\, a_3\, A_{y2}\, (x^2 - a^2) + 4\, a_2\, A_{y1}\, B_x \qquad (4.2.10)$$

and

$$B_x = 14\, x^6 - 18.333\, x^4\, a^2 + 4.333\, x^2\, a^4$$

$$-N_{xy} = 16\, (a_1\, C_x\, C_y + a_2\, D_x\, C_y + a_3\, C_x\, D_y) \qquad (4.2.11)$$

where

$$C_x = x^3 - 3\, x\, a^2$$

$$C_y = y^3 - 3\, y\, b^2$$

$$D_x = 2\, x^7 - 3.666\, x^5\, a^2 + 1.444\, x^3\, a^4$$

$$D_y = 2\, y^7 - 3.666\, y^5\, b^2 + 1.444\, y^3\, b^4$$

Parameters a_i $(i = 3)$ can be calculated, using the following expressions:

$$a_1 = \frac{q\, R_2}{60\, a^6\, \lambda\, (\lambda + \mu_k)} \qquad (4.2.12)$$

$$a_2 = \frac{q\, R_2 - a_1\, a^6\, \lambda\, (11.4\, \lambda + 9.552\, \mu_k)}{a^{10}\, \lambda\, (21.655\, \lambda + 0.972\, \mu_k)}$$

$$a_3 = \frac{q\, R_2 - a_1\, a^6\, \lambda\, (9.552\, \lambda + 11.4\, \mu_k)}{a^{10}\, \lambda^3\, (0.972\, \lambda + 21.655\, \mu_k)}$$

where

$$\lambda = (b/a)^2; \qquad 0.5\, a \le b \le a; \qquad \mu_k = k_1/k_2 = R_2/R_1 \qquad (4.2.13)$$

In the above expressions
a and b are half of the rectangular shell spans in x and y directions, respectively (see Figure 4.2.1);

$k_1 = 1/R_1$ and $k_2 = 1/R_2$ are curvatures and radiuses of curvatures in those directions.

In order to select parameters a_i $(i = 1, 2, 3)$, three co-location points are used. The coordinates of these points are:

$$A\,(0, 0), \qquad B\,(0, 0.9\, b), \qquad C\,(0.9\, a, 0) \qquad (4.2.14)$$

Selection of these points is not random, but depends on the distribution of the normal membrane forces' N_x and N_y in the shell (Figure 4.2.2), which is known from the shell theory and experimental investigations. The first point A is in the center of the shell (at the vertex). Point B is selected near the supports, where the values of N_x are maximal. The forces $N_x = 0$ above the supports $(y = \pm b)$, as the edge elements are infinitely stiff in their plane.

Therefore the deformations $\varepsilon_x = 0$. However, near the supports ($y \approx \pm\, 0.9\, b$) the deformations are small, but non-zero and the forces N_x in this region are rather big. A similar explanation (but in the second direction and for forces N_y) can be given for selecting point C.

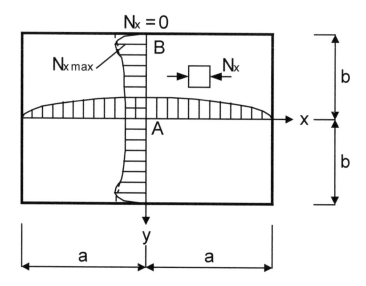

Figure 4.2.2. Diagrams of N_x (A and B—co-location points for x direction).

By rotating Figure 4.2.2 by 90° and replacing B by C a diagram for N_y is obtained.

The membrane horizontal shear forces diagram is represented in Figure 4.2.3.

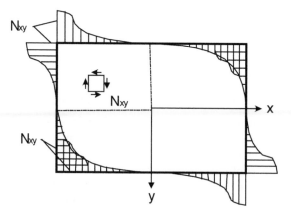

Figure 4.2.3. A diagram of horizontal shear forces N_{xy}.

The main values of forces at the corners can be calculated, using Eq. (4.2.4):

$$N_{m1} = 0.5\,(N_x + N_y) + \sqrt{0.25\,(N_x - N_y)^2 + N_{xy}^2} \qquad (4.2.15)$$

$$N_{m2} = 0.5\,(N_x + N_y) - \sqrt{0.25\,(N_x - N_y)^2 + N_{xy}^2} \qquad (4.2.16)$$

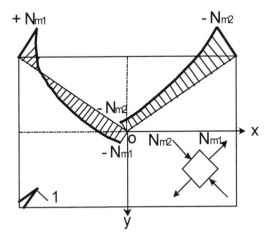

Figure 4.2.4. The main forces' diagram in the directions of the shell diagonals (1 denotes a crack).

The inclination angles of the main forces α_1 and α_2 are:

$$\tan 2\alpha_1 = -\frac{2\,N_{xy}}{N_x - N_y} \qquad (4.2.17)$$

$$\tan 2\alpha_2 = +\frac{2\,N_{xy}}{N_x - N_y} \qquad (4.2.18)$$

4.2.3 Special case—square shell

For a square shell

$$b = a;\ \lambda = 1;\qquad k_1 = k_2 = k;\qquad R_1 = R_2 = R;\qquad \mu_k = 1 \qquad (4.2.19)$$

Therefore, expressions (4.2.12) for parameters a_i take the following form:

$$a_1 = 0.00833\,\frac{q\,R}{a^6};\qquad a_2 = a_3 = 0.0365\,\frac{q\,R}{a^{10}} \qquad (4.2.20)$$

The main forces and their inclination angles can be calculated, taking into account the condition for square shells in direction of their diagonals:

$$N_x = N_y \qquad (4.2.21)$$

Hence

$$N_{m1} = N_x + N_{xy}; \qquad N_{m2} = N_x - N_{xy}; \qquad -\alpha_1 = \alpha_2 = 45° \qquad (4.2.22)$$

It is evident that at the shell corners, when $N_x = 0$, the main forces are equal to the maximum horizontal shear forces:

$$N_{m1} = -N_{m2} = max\, N_{xy} \qquad (4.2.22')$$

Figure 4.2.5 presents the internal membrane forces' diagrams, including the main forces. The co-location points are: (0, 0), (0, 0.9 a) and (0.9 a, 0).

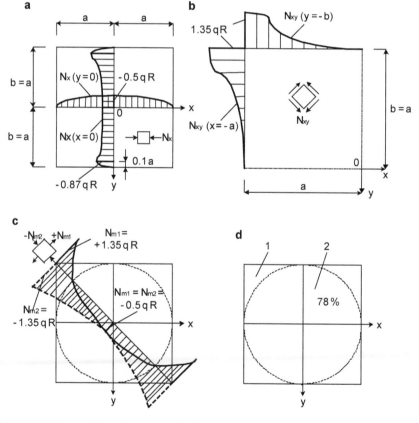

Figure 4.2.5. Internal forces in a square shell:

a—N_x forces at the co-location points; b—horizontal shear forces along the perimeter of the shell; c—main forces; d—compression-tension region (1) and biaxial compression region of the shell (2).

The shell corners are in tension in one direction (perpendicular to the diagonal cracks) and in compression in the other one (parallel to the cracks), as shown in Figure 4.2.5d (region 1). The corresponding area is about 22% of the entire shell surface. In the middle of the region (see region 2 in the figure), the shell is in biaxial compression, and it is at about 78% of the entire shell area. Therefore, such shells are usually made of reinforced concrete in order to use the concrete compressive strength at maximally possible shell area, i.e., it is enough to use minimal reinforcement in this region, like in ordinary compressed elements.

4.2.4 Buckling of convex thin-walled shallow shells

Shells with positive or zero Gaussian curvature can buckle (Figure 4.2.6), as any convex structure, in which thrust forces appear (for example, arch). Usually it is local buckling—a part of the shell changes its shape, i.e., a convex shell becomes concave.

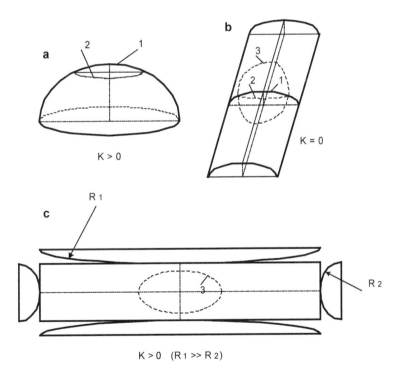

Figure 4.2.6. Buckling of convex RC shells:

a—dome; b—cylindrical shell (with single curvature); c—rectangular shell (with double curvature); 1—shell; 2—buckling shape; 3—buckling zone.

Although buckling of metal shells is well-known, the phenomenon of RC shells became evident just in the sixties of the previous century. Buckling of RC shells has appeared first in relatively small-scale specimens, tested in laboratory conditions (Figure 4.2.7).

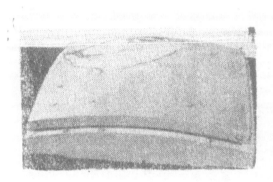

Figure 4.2.7. Buckling of a small-scale RC shell (1.2 × 1.2 m).

The buckling problem becomes especially actual in the following cases:

- for thin-walled shells, when the ratio between the shell thickness and span is rather small;
- if the shell diaphragms have insufficient stiffness in horizontal direction;
- if big elastic-plastic deflections appear due to concrete creep.

In the last decades there has been a tendency to construct thin-walled shells with big spans. For example, the Millennium Dome in London has a diameter of 320 m, the shell of the new terminal at the Charles de Gaulle Airport in Paris was designed to cover a 650 m span in its long direction, the Aqua Park shell in Moscow covered a 70 m span, etc. The second and third structures were made of reinforced concrete and collapsed in buckling on 23/05/2004 and 14/02/2004 respectively. These are two representing examples of buckling that recently appeared in real RC shells. In both cases many people died, i.e., buckling is not just a mistake in design or construction—it is also a social and criminal problem.

The buckling phenomenon is widely investigated in laboratory studies. As is known, critical regions of convex shells become plates and further change their shape to concave (mirror to the initial form, as shown in Figure 4.2.8a). In other words, a convex compressed shell becomes a plate that is in bending and further takes a form of a concave shell that is in tension

(Figure 4.2.8b). It means that during the buckling process the shell changes its static scheme (from compression to tension). For an RC shell, the last case means collapse.

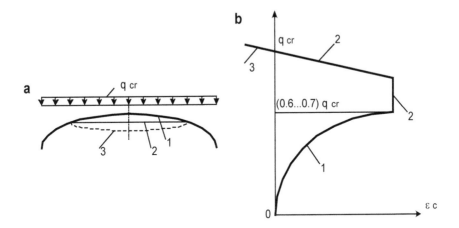

Figure 4.2.8. Buckling as a mirror shape to a convex shell:

a—buckling process (laboratory test); b—concrete deformations at the buckling zone;

1—convex shell; 2—table in bending; 3—concave shell, mirror shape to 1.

In order to prevent buckling of a convex RC shell, the design load F_d that acts on the shell, should not exceed the critical buckling load q_{cr}, i.e.,

$$F_d \leq q_{cr} \tag{4.2.23}$$

where q_{cr} is calculated using the following expression:

$$q_{cr} = 0.2\, E_{c\,red} \left(\frac{h}{R_{max}} \right)^2 k \tag{4.2.24}$$

$E_{c\,red}$ is the reduced modulus of elasticity of concrete that considers concrete creep that decreases the load carrying capacity of statically undetermined structures with time. When the surrounding environment humidity is higher than 40%

$$E_{c\,red} = 0.319\, E_c \tag{4.2.25}$$

else

$$E_{c\,red} = 0.212\, E_c \tag{4.2.26}$$

where h is the shell's thickness;

R_{max} is the maximum value of the radius of curvature for translation shell (the radius of curvature in the other direction is the minimum one R_{min});

k is a coefficient that depends on the ratio R_{max}/R_{min} following Table 4.2.1L.

Table 4.2.1. Coefficients k vs. the ratio R_{max}/R_{min}.

R_{max}/R_{min}	< 1.5	1.5	1.75	2	2.25	2.5	> 2.5
k	1.0	1.17	1.40	1.63	1.79	1.98	2.0

4.2.5 Calculation of shells by local bending moment

The differential equation that considers bending in an element of a translation shell is:

$$D\left(\frac{\partial^4 w}{\partial x^4} + \frac{\partial^4 w}{\partial y^4}\right) - k_x N_x - k_y N_y = q \qquad (4.2.27)$$

where D is the shell cylindrical stiffness:

$$D = \frac{E_c I}{1-v^2} \approx \frac{E_c h^3}{12} \qquad (4.2.28)$$

v is the Poisson's coefficient for concrete ($v = 1/6$, $1 - v^2 \approx 0$)

The terms in the brackets in Eq. (4.2.27) represent the bending state of the shell, and the rest—its membrane state. In case, when the external load is uniformly-distributed on the shell surface, the terms in the brackets have minor influence on the stresses. Therefore, usually these terms can be neglected. However, the influence becomes valuable near the edge elements (where local moments appear due to some restraint of the thin-walled shell in a thick upper belt of the diaphragm). But as the zone of local moments is rather narrow (see Figure 2.2.1), it is possible to assume that these moments in x and y directions are independent, i.e., don't affect each other.

Additionally, moments that act in the same direction from different sides of the shell ($M_{x\,AB}$, $M_{x\,CD}$) have no effect each on the other (Figure 4.2.9a). Near the diaphragms, the membrane forces perpendicular to its plane are equal to zero (Figure 4.2.9b). Therefore, two terms (one of them from the

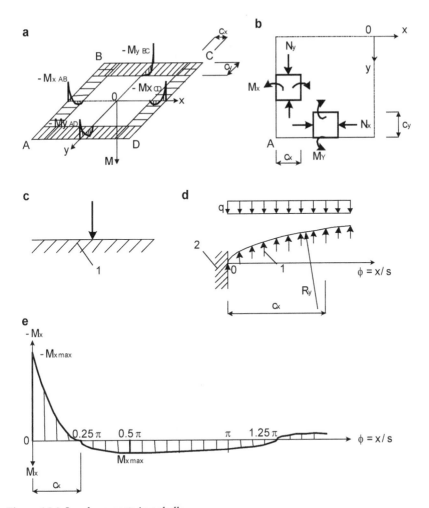

Figure 4.2.9. Local moments in a shell:

a—the local moments zone; b—stress state at the local moments zone;
c—a beam on elastic foundation (1); d—response of a shell similar to the above-mentioned beam (2—restraint); e—local moments diagram.

brackets) in Eq. (4.2.27) become equal to zero. In such a case the equation for *x* direction takes the following form:

$$D\left(\frac{\partial^4 w}{\partial x^4} + \frac{\partial^4 w}{\partial y^4}\right) - k_x N_x - k_y N_y = q \qquad (4.2.29)$$

Substitution of the following expression for N_y into Eq. (4.2.29)

$$N_y = - E_c h \, \varepsilon_y \approx - E_c h \, k_y \, w \qquad (4.2.30)$$

yields to the following equation for the shell bending near the diaphragms:

$$D \frac{\partial^4 w}{\partial x^4} - k_y^2 \, E_c \, hw = q \qquad (4.2.31)$$

where w is the shell deflection.

Dividing this equation by $k_y^2 \, E_c \, h$ yields:

$$\frac{D}{k_y^2 \, E_c \, h} \frac{d^4 w}{d x^4} + w = \frac{q}{k_y^2 \, E_c \, h} \qquad (4.2.32)$$

Let's define

$$\frac{D}{k_y^2 \, E_c \, h} = \frac{s^4}{4} \qquad (4.2.33)$$

and substitute into Eq. (4.2.32). It yields to the following expression:

$$\frac{s^4}{4} \frac{d^4 w}{d x^4} + w = \frac{q}{k_y^2 \, E_c \, h} \qquad (4.2.34)$$

The obtained expression is equivalent to Winkler's differential equation for beams on elastic foundation (Figure 4.2.9a) where s is the elastic foundation characteristic (according to Winkler), i.e., the shell flexibility in vertical direction z. Consequently, the shell near the support behaves as a beam on elastic foundation with characteristic s, and, for example, for x direction (Figure 4.2.9d).

$$s = \sqrt[4]{\frac{4 D}{k_y^2 \, E_c \, h}} = 0.76 \sqrt{R_y \, h} = s_x \qquad (4.2.35)$$

$$s_y = 0.76 \sqrt{R_x \, h}$$

According to the strength of materials' theory

$$M_x = - D \frac{d^2 w}{d x^2} \qquad (4.2.36)$$

hence the solution of differential equation (4.2.35) is:

$$M_x = C_1 \, e^{-\varphi} \cos \varphi + C_2 \, e^{-\varphi} \sin \varphi \qquad (4.2.37)$$

where C_1, C_2 are the integration constants that depend on the boundary conditions (in the given case it is rigid connection);

φ is a relative (non-dimensional) horizontal coordinate (the origin of the coordinate system is located on the edge element as shown in Figure 4.2.9d):

$$\varphi = \frac{x}{s} \qquad (4.2.38)$$

When the shell is rigidly connected to the edge element, or if it is a continuous shell

$$M_x^- = 0.5\, s_x^2\, q\, e^{-\varphi}\, (-\cos\varphi + \sin\varphi) \qquad (4.2.39)$$

Then the maximum negative bending moment for $\varphi = 0$ is (Figure 4.2.9e):

$$M_x^- = -0.5\, q\, s_x^2 = -0.289\, q\, R_y\, h \qquad (4.2.40)$$

This moment becomes zero when

$$\varphi = 0.25\, \pi \qquad (4.2.41)$$

i.e., the local bending moments zone width, c_x, is (see Figure 4.2.9e and Eq. 4.2.38)

$$c_x = x = s\,\phi = 0.76\sqrt{R_y\, h}\cdot 0.25\,\pi = 0.60\sqrt{R_y\, h}\cdot 0.25 \qquad (4.2.42)$$

The positive bending moment

$$M_x = 0.1\, q\, s^2 = 0.058\, q\, R_y\, h = 0.2\, M_x^- \qquad (4.2.43)$$

and it is maximal when $\varphi = 0.5\, \pi$:

$$x = s_x\,\varphi = 0.76\sqrt{R_y\, h}\cdot 0.5\,\pi = 1.2\sqrt{R_y\, h} = 2\, c_x \qquad (4.2.44)$$

The positive moments are small and usually constructive reinforcement is enough to carry them.

4.3 Numerical examples for translation shells

4.3.1 Square shell

Let's treat some numerical examples for a square translation thin-walled shallow RC shell with steel truss diaphragms, to which the shell is rigidly connected (Figure 4.1.2).

Example 1

Given:

$$2\, a = 2\, b = 18 \text{ m}; \quad f_1 = f_2 = f = 1.8 \text{ m}; \quad h = 0.06 \text{ m}; \quad \Delta g_k = 1 \text{ kN/m}^2;$$
$$q_k = 2 \text{ kN/m}^2;$$

$$s_x = s_y = 0.76\sqrt{R_x\, h}$$

It is required to calculate the maximal internal forces in the shell.

Solution

Following the equations, given in Section 4.2.3,

$$R = (a^2 + f^2)/(2f) = (9^2 + 1.8^2)/(2 \cdot 1.8) = 23.4 \text{ m}$$

$$g_k = h\,\gamma = 0.06\ 24 = 1.44 \text{ kN/m}^2$$

$$q_d = 1.4\,(g_k + \Delta g_k) + 1.6\,q_k = 1.4\,(1.44 + 1) + 1.6 \cdot 2 = 6.6 \text{ kN/m}^2$$

The membrane forces are:

$$N_x\,(x = 0, y = 0) = N_y\,(x = 0, y = 0) = -0.5\,q_d\,R = -0.5 \cdot 6.6 \cdot 23.4 = -77.2 \text{ kN/m}$$

At the co-location point, when $x = 0$, $y = 0.9\,a$

$$N_x = -0.87\,q_d\,R = -0.87 \cdot 6.6 \cdot 23.4 = -134.4 \text{ kN/m}$$

At the shell corner ($x = y = a$) the maximum horizontal shear forces are

$$\max N_{xy} = 1.35\,q_d\,R = 1.35 \cdot 6.6 \cdot 23.4 = 208.5 \text{ kN/m}$$

and the maximal main forces at the shell corner are

$$N_{m1} = -N_{m2} = \max N_{xy} = 208.5 \text{ kN/m}$$

$$-\alpha_1 = \alpha_2 = 45°$$

The local moments due to rigid elastic connection of the shell to the diaphragms (see Figure 4.2.6) are:

$$s = 0.76\,(R\,h)^{0.5} = 0.76\,(23.4 \cdot 0.06)^{0.5} = 0.9 \text{ m}$$

$$\max M_x^- = -0.5\,q_d\,s^2 = -0.5 \cdot 6.6 \cdot 0.9^2 = -2.67 \text{ kN m/m}$$

$$c_x\,(M_x^- = 0) = 0.60\,(R\,h)^{0.5} = 0.6\,(23.4 \cdot 0.06)^{0.5} = 0.71 \text{ m}$$

$$\max M_x = 0.1\,q\,s^2 = 0.2\,M_x^- = 0.2 \cdot 2.67 = 0.53 \text{ kN m/m}$$

$$c_x\,(M_x = 0) = 1.2\,(R\,h) = 2\,c_x\,(M_x^- = 0) = 2 \cdot 0.71 = 1.42 \text{ m}$$

Example 2

Given:

For the same shell with $E_c = 23000$ MPa it is required to calculate the critical buckling load for

a) a region with humidity above 40%;
b) a region with humidity below 40%;

Solution

According to equations, given in Section 4.2.4,

$$q_{cr} = 0.2\, E_{c\,red}\,(h/R)^2 = 0.2\, E_{c\,red}\,(0.06/23.4)^2 = 1.315 \cdot 10^{-6}\, E_{c\,red}$$

If the shell is constructed in a region with humidity above 40%, then according to Eq. (4.2.25)

$$E_{c\,red} = 0.319\, E_c = 0.319 \cdot 23000 = 7337\ \text{MPa}$$

$$q_{cr} = 1.315 \cdot 10^{-6}\, 7337 = 9.65 \cdot 10^{-3}\ \text{MPa} = 9.65\ \text{kN/m}^2 > q_d = 6.6\ \text{kN/m}^2 \rightarrow \text{OK}$$

If the shell is constructed in a region with humidity below 40%, then according to Eq. (4.2.26)

$$E_{c\,red} = 0.212\, E_c = 0.212 \cdot 23000 = 4876\ \text{MPa}$$

$$q_{cr} = 1.315 \cdot 10^{-6}\, 4876 = 6.41 \cdot 10^{-3}\ \text{MPa} = 6.41\ \text{kN/m}^2 < q_d = 6.6\ \text{kN/m}^2 \rightarrow$$
$$\text{not OK}$$

i.e., to prevent the shell buckling in the second case higher concrete class (with higher elasticity modulus E_c) should be used.

4.3.2 Rectangular shell

Let's consider a rectangular translation thin-walled shallow RC shell with steel truss diaphragms that are elastic rigid supports for the shell (see Figures 4.2.2–4.2.4).

Example 1

Given:

$$2a = 60\ \text{m}; \quad 2b = 48\ \text{m}; \quad R_1 = 50\ \text{m}; \quad R_2 = 40\ \text{m}; \quad f_1 = 10\ \text{m}; \quad f_2 = 8\ \text{m};$$
$$h = 0.08\ \text{m}; \quad F_d = 4.5\ \text{kN/m}^2 = q$$

It is required to calculate the maximum internal forces in the shell.

Solution

Following the equations from Section 4.2.2,

$$\lambda = (b/a)^2 = (24/30)^2 = 0.64$$
$$\mu_k = k_1/k_2 = R_2/R_1 = 40/50 = 0.8$$

$$a_1 = q\,R_2/[60\,a^6\,\lambda\,(\lambda + \mu_k)] = q\,R_2/[60\,a^6\,0.64\,(0.64 + 0.8)] =$$

$$= 4.5 \cdot 40/(55.25 \cdot 30^6) = 4.47 \cdot 10^{-9}\ \text{kN m}^{-7}$$

$$a_2 = \frac{q\,R_2 - a_1\,a^6\,\lambda\,(11.4\,\lambda + 9.552\,\mu_k)}{a^{10}\,\lambda\,(21.655\,\lambda + 0.972\,\mu_k)} = \frac{q\,R_2}{11.33\,a^{10}} =$$

$$= 4.5 \cdot 40/(11.33 \cdot 30^{10}) = 2.69 \cdot 10^{-14}\ \text{kN m}^{-11}$$

$$a_3 = \frac{q\,R_2 - a_1\,a^6\,\lambda\,(9.552\,\lambda + 11.4\,\mu_k)}{a^{10}\,\lambda^3\,(0.972\,\lambda + 21.665\,\mu_k)} = \frac{q\,R_2}{5.71\,a^{10}} =$$

$$= 4.5 \cdot 40/(5.71 \cdot 30^{10}) = 5.34 \cdot 10^{-14}\ \text{kN m}^{-11}$$

The membrane forces N_x and Ny at the vertex of the shell when $x = y = 0$ can be calculated as follows:

$$A_{x1} = 5\,a^4; \quad A_{y1} = 5\,b^4; \quad A_{x2} = A_{y2} = B_y = B_x = 0;$$

$$N_x = -60\,a_1\,a^4\,b^2 = -1.086\,q\,R_2\,b^2/a^2 = -1.086\,q\,R_2\,\lambda =$$

$$= -1.086 \cdot 4.5 \cdot 40 \cdot 0.64 = -125.1\ \text{kN/m}$$

$$N_y = -60\,a_1\,a^2\,b^4 = -1.086\,q\,R_2\,b^4/a^4 = -1.086\,q\,R_2\,\lambda^2 =$$

$$= -1.086 \cdot 4.5 \cdot 40 \cdot 0.64^2 = -80.03\ \text{kN/m}$$

It should be mentioned that the membrane force N_x that acts in the long direction at the shell vertex is higher than N_y, acting in the short one.

The membrane force N_x at the co-location point $x = 0, y = \pm 0.9\,b$ can be calculated following Eq. (4.2.9):

$$N_x = -2.28\,a_1\,A_{x1}\,b^2 - 2.28\,a_2\,A_{x2}\,b^2 + 4\,a_3\,A_{x1}\,B_y$$

$$A_{x1} = 5\,a^4; \quad A_{x2} = 0; \quad B_y = -1.08\,b^6$$

$$N_x = -11.4\,a_1\,\lambda\,a^6 - 21.6\,a_3\,\lambda^3\,a^{10} = -1.124\,q\,R_2 = -1.124 \cdot 4.5 \cdot 40 = -202.3\ \text{kN/m}$$

The diagram of N_x corresponds to Figure 4.2.2.

The membrane forces N_{xy} at the shell corner, when $x = \pm a, y = \pm b$, can be calculated following Eq. (4.2.11) and Figure 4.2.4. For $x = a, y = b$:

$$C_x = x^3 - 3\,x\,a^2 = -2\,a^3 = -2\,30^3 = -54000\ \text{m}^3$$

$$C_y = y^3 - 3\,y\,b^2 = -2\,b^3 = -2\,24^3 = -27648\ \text{m}^3$$

$$D_x = 2\,x^7 - 3.666\,x^5\,a^2 + 1.444\,x^3\,a^4 = -0.222\,a^7 = -0.222 \cdot 30^7 = -4.85 \cdot 10^9\ \text{m}^7$$

$$D_y = 2\,y^7 - 3.666\,y^5\,b^2 + 1.444\,y^3\,b^4 = -0.222\,b^7 = -0.222 \cdot 24^7 = -1.01 \cdot 10^9\ \text{m}^7$$

$$-N_{xy} = 16\,(a_1\,C_x\,C_y + a_2\,D_x\,C_y + a_3\,C_x\,D_y) =$$

$$= 1.171\,q\,R_2 = 1.171 \cdot 4.5 \cdot 40 = 210.8\ \text{kN/m}$$

The maximum main forces that act at the shell corner are equal to the maximum horizontal shear forces (see Figure 4.2.3):

$$N_{m1} = -N_{m2} = N_{xy} = 210.8 \text{ kN/m}$$

As the forces in x and y directions at the shell corners are equal to zero, then, according to Eqs. (4.2.17) and (4.2.18), the main angles are:

$$-\alpha_1 = \alpha_2 = 45°$$

The local bending moments in x and y directions due to rigid connection of the shell to the diaphragms are different (see Figure 4.2.6a). In x direction

$$s_x = 0.76 \ (R_2 \ h)^{0.5} = 0.76 \ (40 \cdot 0.08)^{0.5} = 1.36 \text{ m}$$

$$s_y = 0.76 \ (R_1 \ h)^{0.5} = 0.76 \ (50 \cdot 0.08)^{0.5} = 1.52 \text{ m}$$

$$\max M_x^- = -0.5 \ q \ s_x^2 = -0.5 \cdot 4.5 \ 1.36^2 = -4.16 \text{ kN m/m}$$

$$c_x \ (M_x^- = 0) = 0.60 \ (R_2 \ h)^{0.5} = 0.6 \ (40 \cdot 0.08)^{0.5} = 1.07 \text{ m}$$

$$\max M_x = 0.1 \ q \ s_x^2 = 0.2 \ M_x^- = 0.2 \cdot 4.16 = 0.83 \text{ kN m/m}$$

$$c_x \ (M_x = \max) = 1.2 \ (R_2 \ h)^{0.5} = 2 \ c_x \ (M_x^- = 0) = 2 \cdot 1.07 = 2.14 \text{ m}$$

In y direction

$$s_y = 0.76 \ (R_1 \ h)^{0.5} = 0.76 \ (50 \cdot 0.08)^{0.5} = 1.52 \text{ m}$$

$$\max M_y^- = -0.5 \ q \ s_y^2 = -0.5 \cdot 4.5 \cdot 1.52^2 = -5.2 \text{ kN m/m}$$

$$c_y \ (M_y^- = 0) = 0.60 \ (R_2 \ h)^{0.5} = 0.6 \ (40 \cdot 0.08)^{0.5} = 1.07 \text{ m}$$

$$\max M_y = 0.1 \ q \ s_y^2 = 0.2 \ M_y^- = 0.2 \cdot 5.2 = 1.04 \text{ kN m/m}$$

$$c_y \ (M_y = \max) = 1.2 \ (R_2 \ h)^{0.5} = 2 \ c_y \ (M_y^- = 0) = 2 \cdot 1.07 = 2.14 \text{ m}$$

Example 2

Given:

For the same shell with $E_c = 23000$ MPa it is required to calculate the critical buckling load for a region with humidity above 40%.

Solution

Following Eq. (4.2.24) and Table 4.2.1, for the given shell

$$R_{max}/R_{min} = 50/40 = 1.25, k = 1$$

$$q_{cr} = 0.2 \ E_{c \ red} \ (h/R_{max})^2 = 0.2 \ E_{c \ red} \ (0.08/50)^2 = 0.512 \ 10^{-6} \ E_{c \ red}$$

If the shell is constructed in a region with humidity above 40%, then according to Eq. (4.2.25)

$$E_{c\,red} = 0.319\ E_c = 0.319 \cdot 23000 = 7337\ \text{MPa}$$

$$q_{cr} = 0.512 \cdot 10^{-6} \cdot 7337 = 3.76 \cdot 10^{-3}\ \text{MPa} = 3.76\ \text{kN/m}^2 < q = 4.5\ \text{kN/m}^2 \rightarrow \text{not OK}$$

i.e., the shell may buckle. To prevent buckling, higher concrete class (with higher modulus of elasticity E_c) should be used, or alternatively, the shell thickness should be increased to 10 cm.

4.3.3 Calculation of N_{xy} forces in the upper truss belt

Given that the maximal horizontal shear force N_{xy} = 193.68 KN/m. Length of each field between the truss joints is equal to 3.5 m. Numerical values of a_1, a_2, a_3, C_y, D_y are also known. It is possible to calculate the forces N_{xy}, transferred to each joint of the truss (as it was shown before, the forces N_{xy} at the center of the truss are equal to zero).

It is required to calculate the axial forces in the joints of the truss upper belt F_i ($i = 1 \dots 5$) (Figure 4.3.1).

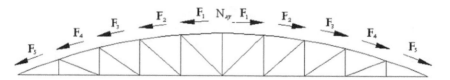

Figure 4.3.1. Forces N_{xy} transferred from the shell to the truss.

Solution

$$N_{xy} = 16 \left(\alpha_1 C_x C_y + \alpha_2 C_x C_y + \alpha_3 C_x C_y \right)$$

$$C_x = x^3 - 3xa^2$$

$$D_x = 2\,x^7 - 3.666\ x^5a^2 + 1.444\ x^3a^4$$

$x_1 = 3.5$ m	$\rightarrow$	$C_x = -3172.75,$	$D_x = 5.23 \cdot 10^6$
$x_2 = 7$ m	$\rightarrow$	$C_x = -6088.25,$	$D_x = 2.92 \cdot 10^7$
$x_3 = 10.5$ m	$\rightarrow$	$C_x = -8489.25,$	$D_x = 4.16 \cdot 10^7$
$x_4 = 14$ m	$\rightarrow$	$C_x = -10118.5,$	$D_x = -2.1 \cdot 10^7$
$x_5 = 17.5$ m	$\rightarrow$	$C_x = -10718.8,$	$D_x = -1.1 \cdot 10^8$

Substituting C_x and D_x into N_{xy} yields:

$$N_{xy\,1} = 37.27 \text{ kN/m}$$
$$N_{xy\,2} = 61.38 \text{ kN/m}$$
$$N_{xy\,3} = 85.13 \text{ kN/m}$$
$$N_{xy\,4} = 139 \text{ kN/m}$$
$$N_{xy\,5} = N_{xy} = 193.68 \text{ kN/m}$$

Calculation of forces at the truss joints:

$$F_1 = x \cdot N_{xy\,1}/2 = 3.5 \cdot 37.27/2 = 65.22 \text{ kN}$$
$$F_2 = x \cdot (N_{xy\,1} + N_{xy\,2})/2 = 3.5\,(37.27 + 61.38)/2 = 172.62 \text{ kN}$$
$$F_3 = x \cdot (N_{xy\,2} + N_{xy\,3})/2 = 3.5\,(61.38 + 85.13)/2 = 256.39 \text{ kN}$$
$$F_4 = x \cdot (N_{xy\,3} + N_{xy\,4})/2 = 3.5\,(85.13 + 139)/2 = 392.22 \text{ kN}$$
$$F_5 = x \cdot (N_{xy\,4} + N_{xy\,5})/2 = 3.5\,(139 + 193.68)/2 = 582.19 \text{ kN}$$

5

Long Convex Cylindrical Shells

5.1 Reinforced concrete shell structures

5.1.1 General

A general view of a typical RC cylindrical shell is shown in Figure 2.1.2b in Section 2. Such shells consist of three main structural elements:

- a thin-walled translation shallow shell;
- edge elements in the long direction (types of these elements are shown in Figure 2.1.4);
- diaphragms in the short (curved) direction (types of diaphragms are shown in Figure 2.1.3).

In long cylindrical shells, the ratio between the spans should satisfy the following requirement:

$$\frac{l_1}{l_2} = \frac{a}{b} \geq 2 \tag{5.1.1}$$

Figure 5.1.1 presents such a shell and the main internal forces that act in it under a uniform distributed external load:

- longitudinal membrane force in the long shell direction, N_x;
- horizontal shear membrane force, N_{xy};
- transverse bending moment in the short (curved) direction of the shell, M_y.

In this case, it is possible to calculate the main forces and their directions, using the following simple expressions:

$$N_{m1} = \frac{N_x}{2} + \sqrt{\frac{N_x^2}{4} + N_{xy}^2} \tag{5.1.2}$$

$$N_{m2} = \frac{N_x}{2} - \sqrt{\frac{N_x^2}{4} + N_{xy}^2} \qquad (5.1.3)$$

$$- \tan 2\alpha_1 = \tan 2\alpha_2 = \frac{2\,N_{xy}}{N_x} \qquad (5.1.4)$$

In addition to the transverse moments, local bending moments appear in the shell near the diaphragms, similar to translation shells.

The reinforcement scheme, according to calculations (and not following constructive requirements!) and corresponding to all above-mentioned internal forces, is shown in Figure 2.2.1b. The rest of internal forces (N_y, M_x) in long cylindrical shells are rather small and are usually neglected (just constructive reinforcement is used).

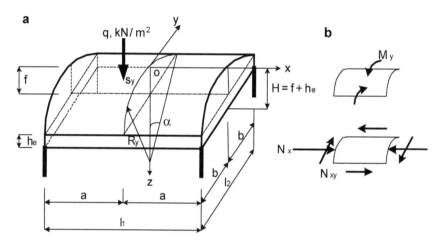

Figure 5.1.1. A cylindrical shell: a—general view; b—main internal forces.

5.1.2 Constructive requirements to shell dimensions

Usually long cylindrical thin-walled shallow shells are used. Therefore, the shell thickness

$$h = \left(\frac{1}{100} \cdots \frac{1}{300}\right) l_2 \geq 5 \text{ cm} \qquad (5.1.5)$$

The rise of the shell itself is

$$f = \left(\frac{1}{6} \cdots \frac{1}{8}\right) l_2 \qquad (5.1.6)$$

and the entire shell height

$$H = \left(\frac{1}{10} \cdots \frac{1}{15}\right) l_1 \tag{5.1.7}$$

The rectangular section edge elements' height

$$h_e = \left(\frac{1}{20} \cdots \frac{1}{30}\right) l_1$$

and their width

$$b_e = (0.3 \ldots 0.4) h_e \tag{5.1.8}$$

Openings are allowed in the upper part of the shell (Figure 5.1.2) and their width

$$c \leq (0.25 \ldots 0.33) l_2 \tag{5.1.9}$$

At the same time, a strengthening ribs should be used each 2 … 3 m around the openings.

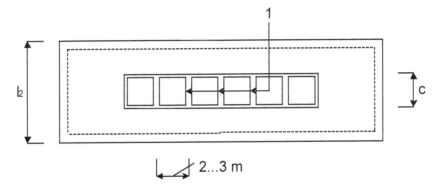

Figure 5.1.2. An upper view on openings in a shell (1—strengthening ribs).

The longitudinal reinforcement is usually located in the bottom zone of the edge element, in order to achieve maximum lever of the internal forces. This reinforcement may be shortened according to the bending moments' diagram, but at least 30% should be extended up to the supports. In the shell itself constructive longitudinal reinforcement is provided (without calculations) with $\rho_{min} = 0.2\%$. In the perpendicular direction, reinforcement, corresponding to the transverse bending moments, is used. The reinforcement bars in both directions should be tied at all mesh nodes, or alternatively, welded reinforcement mesh can be used. As the shell is rigidly connected to the edge element, $\varnothing$ (6…8), @ ≤ 20 reinforcement bars are provided perpendicular to the edge element. The length of these bars is according to calculation by local moments, but at least $0.1\ l_2$.

5.1.3 A cylindrical shell of the 2E terminal at the Charles de Gaulle Airport

Let's discuss the covering structure of the 2E terminal at the Charles de Gaulle Airport in Paris (Figure 5.1.3). The roof consists of a number of convex cylindrical shells that will be explained in details later. The total dimensions of the building are 650 × 30 m (upper view). The shape of the shell is a part of an ellipsoid with main axes dimensions of 29 m and 17 m. The shell span between the supports (columns) is $l_2 = 26$ m (Figure 5.1.4a). The radius of curvature at the vertex is $R_y = 24.7$ m. The length of the shell is $l_1 = 650$ m. The structure consists of 10 parts, separated by glass straps. Each part is an assembly of pre-cast shell elements (4 m in width and 26 m in length) that are simple supported on rows of columns.

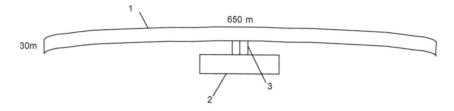

Figure 5.1.3. The 2E terminal at the Charles de Gaulle Airport in Paris (upper view):

1—the waiting area before boarding; 2—arrival hall at the main terminal; 3—connecting building.

Each of the shells (Figure 5.1.4b) consists of two external and one middle pre-cast elements with a thickness of 30 cm. The elements are connected by concrete, casted at the site (Figure 5.1.4c). These elements include the edge elements in a form near to triangular (Figure 5.1.4d). As the out-of-plane (horizontal) stiffness of the edge elements is rather low, the external parts of the shell are strengthened by external diaphragms, made of steel pipes (Figure 5.1.4e).

The design load that acts on the shell is 13 kN/m², including the snow (the part of self-weight, dead and service loads is about 12 kN/m²). The shell was in service for more than two years, i.e., there is an instant load influence and the long-term one too. The shell is thin-walled, because the ratio between the thickness and the span is $h/l_2 \approx 1/100$. Considering the influence of long-term loads in such shells is very important because of their geometrical non-linearity and due to stresses' state too. Buckling analysis of this shell will be presented in the Section 5.4.3.

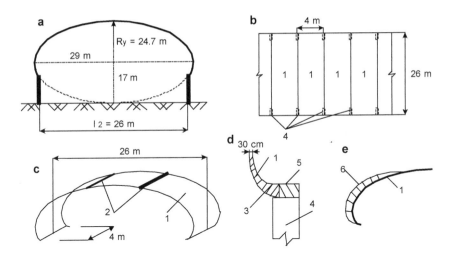

Figure 5.1.4. 2E terminal cylindrical shell:

a—the shell section; b—upper view; c—elliptic part; d—connection between the part and column; e—diaphragms; 1—part of the shell 4 × 26 m; 2—joint between the pre-cast elements; 3—edge element; 4—columns; 5—beam above the columns; 6—external diaphragms (steel pipes), anchored in the shell.

5.2 Calculation of internal forces in a long cylindrical shell

5.2.1 Cylindrical shell with infinitely stiff edge element

An equation of a cylindrical shell in a system of coordinates, shown in Figure 5.1.1, corresponds to a quadratic parabola:

$$z = f\left(\frac{y}{b}\right)^2 \tag{5.2.1}$$

i.e., the surface is translation and has a single curvature. Therefore the curvatures of the shell are:

$$k_x = \frac{\partial^2 z}{\partial x^2} = 0; \ k_y = \frac{\partial^2 z}{\partial y^2} = 2\frac{f}{b^2} = \text{const}; \ k_{xy} = \frac{\partial^2 z}{\partial x \, \partial y} = 0 \tag{5.2.2}$$

Correspondingly, the radius of curvature in the transverse direction of the shell (see Figure 5.1.1) is:

$$R_y = \frac{1}{k_y} = \frac{b^2}{2f} \tag{5.2.3}$$

If the edge element of the shell is ideal, i.e., infinitely stiff in its plane (vertical direction) and infinitely flexible out-of-the plane (horizontal direction), it is possible to calculate the internal membrane forces in the shell (longitudinal compression force N_x and horizontal shear force N_{xy}, shown in Figure 5.1.1), using Eq. (4.2.3). Taking into account Eqs. (5.2.2) and (5.2.3) as well as considering that $N_y = \dfrac{d^2\varphi}{dx^2}$ yields:

$$\frac{d^2\varphi}{dx^2} = -q\,R_y \tag{5.2.4}$$

where $\varphi(x, y)$ is the stresses' function (see Chapter 4, Eq. 4.2.8) that was selected according to the following boundary conditions:

$$\text{if } x = \pm a, \quad \text{or/and } y = \pm b, \quad N_x = 0 \tag{5.2.5}$$

The stresses' function can be selected in the following form:

$$\varphi\,(x, y) = a_1\,A_{x1}\,A_{y1} + a_2\,A_{x2}\,A_{y1} \tag{5.2.6}$$

where

$$A_{x1} = x^4 - 6\,x^2\,a^2 + 5\,a^4 \tag{5.2.7}$$

$$A_{y1} = y^4 - 6\,y^2\,b^2 + 5\,b^4$$

$$A_{x2} = x^6 - 2.5\,x^4\,a^2 + 1.5\,a^6$$

$$A_{y2} = y^6 - 2.5\,y^4\,b^2 + 1.5\,b^6$$

At the co-location points $(0, 0)$ and $(0, 0.5\,b)$

$$a_1 = \frac{q\,R_y}{60\,a^2\,b^4}\,; \quad a_2 = \frac{q\,R_y}{16.45\,a^4\,b^6} \tag{5.2.8}$$

The internal membrane forces can be calculated according to the following expressions:

$$N_x = \frac{d^2\varphi}{dy^2} = 12\,a_1\,A_{x1}\,(y^2 - b^2) + 30\,a_2\,A_{x2}\,y^2\,(y^2 - b^2) \tag{5.2.9}$$

$$N_{xy} = -\frac{d^2\varphi}{dx\,dy} = -16\,a_1\,C_x\,C_y - 4\,a_2\,D_x\,D_y \tag{5.2.10}$$

where

$$C_x = x^3 - 3\,x\,a^2 = x\,(x^2 - 3\,a^2) \tag{5.2.11}$$

$$C_y = y^3 - 3\,y\,b^2 = y\,(y^2 - 3\,b^2)$$

$$D_x = 3\,x^5 - 5\,x^3\,a^2 = x^3\,(3\,x^2 - 5\,a^2)$$

$$D_y = 3\,y^5 - 5\,y^3\,b^2 = y^3\,(3\,y^2 - 5\,b^2)$$

Substitution of (5.2.11) into (5.2.10) yields:

$$N_{xy} = -16\, q\, R_y\, x\, (x^2 - 3\, a^2)\, y\, (y^2 - 3\, b^2)/(60\, a^2\, b^4) -$$
$$- 4\, q\, R_y\, x^3\, (3\, x^2 - 5\, a^2)\, y^3\, (3\, y^2 - 5\, b^2)/(16.45\, a^4\, b^6)$$

At the diaphragm, where $x = a$

$$N_{xy} = 8\, q\, R_y\, a\, y\, (y^2 - 3\, b^2)/(15\, b^4) + 8\, q\, R_y\, a\, y^3\, (3\, y^2 - 5\, b^2)/(16.45\, b^6)$$

It is easy to prove that the integral of forces N_{xy} (that equals to the thrust force) if y changes from zero to b

$$H = 1.03\, q\, R_y\, a$$

5.2.2 Membrane forces diagrams

At the shell center, when $x = y = 0$

$$N_x = q\, R_y\, a^2/b^2 \qquad\qquad (5.2.12)$$

For $x = 0$ and $y = 0.5\, b$

$$N_x = 1.26\, q\, R_y\, a^2/b^2 \qquad\qquad (5.2.13)$$

According to the boundary condition (5.2.5) along the shell perimeter

$$N_x = 0 \qquad\qquad (5.2.14)$$

The diagram of forces N_x is given in Figure 5.2.1a.

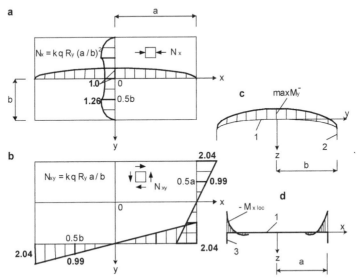

Figure 5.2.1. Forces' diagrams: a—N_x; b—N_{xy}; c—M_y; d—$M_{x\,loc}$;

k—coefficient, which values are given on the forces' diagrams; 1—shell; 2—edge elements; 3—diaphragms.

The maximal horizontal shear membrane forces at the corners are (Figure 5.2.1b):

$$N_{xy} = 2.04 \, q \, R_y \, a/b \tag{5.2.15}$$

5.2.3 Bending moments in the transverse direction of the shell

The differential equation that describes the bending state in the transverse direction of a long cylindrical shell is:

$$\frac{h^3 \, R_y}{12} \frac{d^4 w}{dy^4} - w = \frac{q \, R_y^2}{E_c \, h} \tag{5.2.16}$$

where $w = w(x, y)$ is the vertical displacement of the shell.

The solution of this equation is similar to that for bending in translation shells. The final expression for the transverse bending moment is:

$$M_y = - 3 \, h^2 \, R_y \, (4 \, a_1 + 25 \, a_2 \, x^2 \, y^2) \, (x^2 - a^2) \, (y^2 - b^2) \tag{5.2.17}$$

where a_1 and a_2 are obtained according to expressions (5.2.8).

At the middle of the shell, when $x = y = 0$

$$\max M_y^- = - 0.2 \, q \, h^2 \, R_y^2 / b^2 \tag{5.2.18}$$

At the long edges of the shell, where $x = 0, y = \pm b$

$$M_y = 0 \tag{5.2.19}$$

When $x = 0$ and $y | \neq b$ the value of M_y is always negative:

$$M_y < 0 \tag{5.2.20}$$

i.e., the diagram of transverse bending moments M_y in general case has always been a single sign (Figure 5.2.1c).

5.2.4 Local bending near the diaphragms

The local bending moment $M_{x \, loc}$ appears near the diaphragms and calculated using the same method like for translation shells (see Section 4.2.5 and Figure 4.2.9e in Chapter 4). The only difference is that in translation shells these moments appear along the perimeter (see Figure 4.2.9a), and in cylindrical ones—just near the diaphragms in the short direction of the shell (Figure 5.2.1d). The corresponding equations are (4.2.40) ... (4.2.45).

5.2.5 Calculating the shell buckling capacity

Calculation of cylindrical shell buckling capacity is carried out considering long-term concrete creep and taking into account the fact that edge elements of the shell are not able to take forces that act perpendicular to their planes. The buckling shape of cylindrical shells is similar to that of arcs. The main difference between shells and arcs is that in convex shells the buckling area is some zone similar to ellipsoid (see Figure 4.2.7), whereas in arcs buckling appears as a line.

The method for calculating buckling of shells with zero Gaussian curvature ($K = 0$) depends on the shell and on its boundary conditions. When speaking about analysis of a structure that comes to failure in buckling, all possible failure schemes that are able to withstand the load and the given surrounding conditions should be checked and a critical load for each scheme should be found. In case of an RC shell, the long-term concrete creep that may reduce the load carrying capacity of statically undetermined structures should be considered additionally. It is clear that the minimal value of all critical forces corresponds to the structure's real buckling capacity. Selecting appropriate expressions for calculating the buckling capacity of a long cylindrical convex RC shell with zero Gaussian curvature depends on parameter m:

$$m = 1.76\alpha_0 \sqrt[8]{(1-\mu^2)\frac{R_y^6}{l_2^6}}, \quad \text{rad} \tag{5.2.21}$$

Considering that $(1-\mu^2)^{1/8} \approx 1$ for Poisson coefficient $\mu \approx 0.2$ in case of concrete yields:

$$m = 1.76\alpha_0 \sqrt[4]{\frac{R_y^3}{l_2^3}}, \quad \text{rad} \tag{5.2.22}$$

If $m \leq 1$ the critical buckling load for a shell is calculated as follows:

$$q_{cr} = E_c \frac{h}{R_y}\left[0.857\frac{h^2}{s_y^2} + 0.101\frac{s_y^2 h^4}{R_y^2 l_2^4}\right] \tag{5.2.23}$$

else

$$q_{cr} = 0.882 E_c \frac{h}{l_2}\left(\frac{h}{R_y}\right)^{3/2} \tag{5.2.24}$$

where s_y is the length of the shell's transverse section arch:

$$s_y = 2 R_y \alpha_0 \tag{5.2.25}$$

α_0 is a half of the central angle that corresponds to the arch (see Figure 5.1.1a):

$$\alpha_0 = \text{arc sin } (b/R_y), \quad \text{rad} \tag{5.2.26}$$

The modulus of elasticity of the shell's concrete, E_c, reduces with time due to concrete creep, and for an infinitely long time period it can be obtained as follows:

- for humidity higher than 40%

$$E_{c.red} = 0.319 \, E_c \tag{5.2.27}$$

- for humidity lower than 40%

$$E_{c.red} = 0.212 \, E_c \tag{5.2.28}$$

5.3 Calculating a long cylindrical shell as a simple supported beam

5.3.1 General

A long cylindrical shell $(l_1/l_2 \geq 2)$ that is simple supported on columns at the corners can be considered as a beam with a span l_1 in the long direction. A system of typical cracks that appear in such shell is shown in Figure 5.3.1. The cracks include:

- vertical cracks in the edge elements as a result of the shell bending, similar to a simple beam (see pos. 1 in Figure 5.3.1);
- transverse crack along the shell vertex due to negative transverse bending moments in the shell itself (see pos. 2 in Figure 5.3.1);
- diagonal cracks at the shell corners due to the main tensile stresses (see pos. 3 in Figure 5.3.1);
- transverse cracks along the shell sides due to negative local bending moments in the shell (see pos. 4 in Figure 5.3.1).

The upper picture in Figure 5.3.1 presents the failure mode of an RC shell model. The dimensions of the model are 0.6 × 3.6 m and it is a 1:5 model of a 3 × 18 m real shell-panel. The model was supported at the corners and it behaved as a simple supported beam. The edge elements were rather flexible in the vertical direction and cracked due to bending under uniformly distributed loading (see pos. 5 in Figure 5.3.1). This crack appeared due to reinforcement bars' yielding.

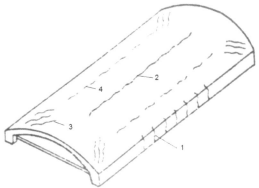

Figure 5.3.1. Laboratory testing of a long cylindrical shell model: the model's failure and a system of expected cracks 1–4 (explained above).

5.3.2 Cylindrical shell with finite stiffness of edge elements

Let's calculate a long cylindrical shell in its long direction, using a static scheme of a simple supported beam. The section of the shell includes an arch of the shell itself and two rectangular edge elements (Figure 5.3.2). These elements are located in tensile zone and include longitudinal reinforcement A_s. The long span direction is usually big (l_1 is up to 30 m) and reinforcement in this direction can be pre-stressed. A basic assumption for calculating long shells as beams is that there are no in-plane deformations of the shell section in transverse direction l_2. As it is evident from available experimental data, in real shells such deformations exist, therefore their load carrying capacity is 10 … 20% lower than the calculated one.

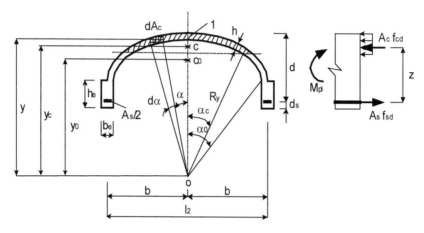

Figure 5.3.2. Transverse section of the shell, calculated using a simple supported beam static scheme (1—compressed zone).

The section equilibrium equations in the ultimate limit state are:

$$A_s f_{sd} = A_c f_{cd} \tag{5.3.1}$$

$$M_{pl} = A_c f_{cd}\, z = A_s f_{sd}\, z \tag{5.3.2}$$

where A_c is the section's compressed zone area (hatched zone in Figure 5.3.2);

M_{pl} is the moment, carried by the section at plastic failure, due to reinforcing steel yielding.

Solution of the above equations depends on the aim of the calculation:

a) for calculating the value of M_{pl} and checking if it is equal or higher than the external forces' moment, M_d,

$$M_{pl} \geq M_d \tag{5.3.3}$$

b) for calculating the reinforcement section A_s, required to carry M_d.

a) Checking the section

The compressed zone area, A_c, is calculated as follows:

$$A_c = 2 \int_0^{\alpha c} h \, R_y \, d\alpha = 2 h R_y \, \alpha_c \qquad (5.3.4)$$

where α_c is a half of the central angle, corresponding to the compressed section zone.

From Eq. (5.3.1) follows:

$$\alpha_c = \frac{A_s \, f_{sd}}{2 \, h \, R_y \, f_{cd}} \qquad (5.3.5)$$

The moment carried by the section M_{pl} is calculated using Eq. (5.3.2), where the lever of the internal forces

$$z = y_c - (R_y - d) \qquad (5.3.6)$$

and y_c is a distance between the center of the arch section's curve (point o in Figure 5.3.2) and the center of gravity of its compressed zone (point c):

$$y_c = S_{co}/A_c \qquad (5.3.7)$$

S_{co} is the static moment of the compressed zone area relative to point o:

$$S_{co} = 2 \int_0^{\alpha c} y \, dA_c \qquad (5.3.8)$$

y is a distance between o and the differential area dA_c:

$$y = R_y \cos \alpha \qquad (5.3.9)$$

$$dA_c = h \, R_y \, d\alpha \qquad (5.3.10)$$

$$S_{co} = 2 \, h \, R_y^2 \sin \alpha_c \qquad (5.3.11)$$

$$y_c = R_y \frac{\sin \alpha_c}{\alpha_c} \qquad (5.3.12)$$

$$z = R_y \frac{\sin \alpha_c}{\alpha_c} - (R_y - d) \qquad (5.3.13)$$

$$M_{pl} = 2 h R_y \, \alpha_c \, f_{cd} \left[R_y \frac{\sin \alpha_c}{\alpha_c} - (R_y - d) \right] \qquad (5.3.14)$$

It should be checked, if the requirement (5.3.3) is satisfied for this value of M_{pl}, and if the ratio

$$S_c/S_0 \le 0.64 \qquad (5.3.15)$$

where S_0 is a static moment of the equivalent section of the entire shell and edge elements, relative to the center of gravity of the tensile reinforcement A_s (see Figure 5.3.2);

S_c—same for the compressed section zone:

$$S_0 = A_{0\,sh}\,[y_0 - (R_y - d)] + A_{0\,e}\,(h_e - d_s)/2 \tag{5.3.16}$$

$$A_{0\,sh} = 2\,R_y\,h\,\alpha_0; \quad y_0 = R_y\,\sin\alpha_0/\alpha_0; \quad A_{0\,e} = 2\,b_e\,(h_e - d_s)$$

$$S_0 = 2\,R_y\,h\,\alpha_0\,[R_y\,\sin\alpha_0/\alpha_0 - (R_y - d)] + 2\,b_e\,(h_e - d_s)\,(h_e - d_s)/2 =$$

$$= 2\,R_y\,h\,\alpha_0\,[R_y\,\sin\alpha_0/\alpha_0 - (R_y - d)] + b_e\,(h_e - d_s)^2$$

$$S_c = A_c\,z = 2\,h\,R_y\,\alpha_c\,[R_y\,\frac{\sin\alpha_c}{\alpha_c} - (R_y - d)] \tag{5.3.17}$$

b) Calculating the reinforcement section

According to the first expression in Eq. (5.3.2) for

$$M_{pl} = M_d \tag{5.3.18}$$

and using Eq. (5.3.14),

$$M_d = 2\,h\,R_y\,\alpha_c\,f_{cd}\left[R_y\,\frac{\mathrm{Sin}\,\alpha_c}{\alpha_c} - (R_y - d)\right] \tag{5.3.19}$$

Defining non-dimensional values

$$A = M_d/(2\,h\,R_y^2\,f_{cd}) \quad \text{and} \quad B = 1 - d/R_y \tag{5.3.20}$$

yields the following transcendent equation:

$$\sin\alpha_c - B\,\alpha_c = A \tag{5.3.21}$$

Solution of this equation is approximate. As it follows from available experimental data [1, 7], generally, the angle α_c is rather small, therefore it is possible to start the iterative process from the condition:

$$\sin\alpha_c \approx \alpha_c; \quad \alpha_c \approx M_d/(2\,h\,d\,R_y\,f_{cd}) \tag{5.3.22}$$

After the value of α_c is calculated, the following condition should be satisfied:

$$z \le 0.95\,d \tag{5.3.23}$$

Further, the forces' equilibrium equation (5.3.1) is used, considering (5.3.23), for calculating A_s.

5.4 Numerical examples for calculating long cylindrical shells

5.4.1 A cylindrical shell as a simple supported beam: calculating the reinforcement section area in the edge element

Given a long cylindrical RC shell, that has edge elements with finite stiffness. Concrete class is C 30 and ribbed bars are used as reinforcement. Known:

$l_1 = 2 a = 20$ m; $\quad l_2 = 2 b = 5$ m; $\quad h = 0.07$ m; $\quad f = 1$ m; $\quad h_e = 0.5$ m; $b_e = 0.15$ m; $\quad d_s = 3$ cm; $\quad \Delta g_k = 0.3$ kN/m²; $\quad q_k = 0.5$ kN/m².

It is required to calculate A_s.

Solution

$$d = f + h_e - d_s = 1 + 0.5 - 0.03 = 1.47 \text{ m}$$

$$R_y = (b^2 + f^2)/(2 f) = (2.5^2 + 1^2)/(2 \cdot 1) = 3.625 \text{ m}$$

$$\alpha_0 = \text{arc sin } (b/R_y) = \text{arc sin } (2.5/3.625) = 43.6° = 0.76 \text{ rad}$$

$$\sin \alpha_0 = 0.6896$$

The entire shell section area (including the edge elements) is:

$$A_g = 2 (R_y h \alpha_0 + h_e b_e) = 2 (3.625 \cdot 0.07 \cdot 0.76 + 0.5 \cdot 0.15) = 0.5357 \text{ m}^2$$

$$g_k = A_g \cdot 24 = 0.5357 \cdot 24 = 12.9 \text{ kN/m}$$

$$F_d = 1.4 (g_k + \Delta g_k l_2) + 1.6 q_k l_2 = 1.4 (12.9 + 0.3 \cdot 5) + 1.6 \cdot 0.5 \cdot 5 = 24.2 \text{ kN/m}$$

$$M_d = F_d l_1^2/8 = 24.2 \cdot 20^2/8 = 1210 \text{ kN m}$$

$$A = M_d/(2 h R_y^2 f_{cd}) = 1210 \cdot 10^6/(2 \cdot 70 \cdot 3625^2 \cdot 10.7) = 0.0615$$

$$B = 1 - d/R_y = 1 - 1.47/3.625 = 0.595$$

The transcendent equation has the following form:

$$\sin \alpha_c - 0.595 \, \alpha_c = 0.0615$$

The final solution (after a number of iterations) is:

$$\alpha_c = 0.156 \text{ rad} = 8.94°; \sin \alpha_c = 0.155$$

$$y_c = R_y \frac{\sin \alpha_c}{\alpha_c} = 3.625 \cdot 0.155/0.156 = 3.602 \text{ m}$$

$$z = R_y \frac{\sin \alpha_c}{\alpha_c} - (R_y - d) = y_c - (R_y - d) = 3.602 - (3.625 - 1.47) = 1.447 \text{ m}$$

$$0.95\,d = 0.95 \cdot 1.47 = 1.3965\text{ m} < z = 1.447\text{ m} \rightarrow \text{not OK}$$

$$z = z_{max} = 0.95\,d = 1.3965\text{ m}$$

According to z_{max} a new value of α_c is calculated and defined as α_{c1}:

$$z_{max} = R_y \frac{\sin \alpha_{c1}}{\alpha_{c1}} - (R_y - d)$$

$$\sin \alpha_{c1}/\alpha_{c1} = [z_{max} + (R_y - d)]/R_y = [1.3965 + (3.625 - 1.47)]/3.625 = 0.979$$

$$\frac{\sin \alpha_{c1}}{\alpha_{c1}} = 0.959;\ \alpha_{c1} \approx 0.3498\text{ rad} = 20.04° < \alpha_0 = 43.6°$$

As α_{c1} is a minimal angle that corresponds to the section compressed zone depth, it is possible to skip checking if requirement (5.3.15) is satisfied.

$$A_s f_{sd} = A_c f_{cd};\ A_s f_{sd} = 2\,h\,R_y\,\alpha_c f_{cd};\ \alpha_c = \alpha_{c1}$$

$$A_s = 2\,h\,R_y\,\alpha_{c1} f_{cd}/f_{sd} = 2 \cdot 70 \cdot 3625 \cdot 0.3498 \cdot 10.7/350 = 5427\text{ mm}^2$$

$$A_{s\,min} = 0.0015\,A_0 = 0.0015\,(A_g - 2\,b_e\,d_s) =$$

$$= 0.0015\,(0.5357 - 2 \cdot 0.15 \cdot 0.03) = 7.9 \cdot 10^{-4}\text{ m}^2 = 7.9\text{ cm}^2 < A_s = 54.27\text{ cm}^2 \rightarrow \text{OK}$$

5.4.2 Cylindrical shell with infinitely stiff edge element

Given a long cylindrical RC shell with infinitely stiff edge elements (for example, load bearing walls). Concrete class C 25 and ribbed bars reinforcement are used.

Known:

$$l_1 = 2\,a = 24\text{ m};\quad l_2 = 2\,b = 6\text{ m};\quad h = 0.1\text{ m};\quad f = 1.2\text{ m};$$
$$\Delta g_k = 1.0\text{ kN/m}^2;\quad q_k = 2.5\text{ kN/m}^2$$

It is required to calculate the internal forces in the shell.

Solution

$$R_y = (b^2 + f^2)/(2f) = (3^2 + 1.2^2)/(2 \cdot 1.2) = 4.35\text{ m}$$

$$\alpha_0 = \text{arc sin } (b/R_y) = \text{arc sin } (3/4.35) = 43.6°$$

$$g_k = 0.1 \cdot 24 = 2.4\text{ kN/m}^2$$

$$F_d = 1.4\,(g_k + \Delta g_k) + 1.6\,q_k = 1.4\,(2.4 + 1.0) + 1.6 \cdot 2.5 = 8.8\text{ kN/m}^2$$

Let's calculate the membrane transverse forces N_x at the co-location point:

When $x = y = 0$, according to Eq. (5.2.12) for $q = F_d$

$$N_x = F_d R_y a^2/b^2 = 8.8 \cdot 4.35 \cdot 12^2/3^2 = 612.5 \text{ kN/m}$$

When $x = 0$ and $y = 0.5\,b$

$$N_x = 1.26\, F_d R_y a^2/b^2 = 434.9 \cdot 1.26/0.71 = 771.8 \text{ kN/m}$$

The diagram of N_x is presented in Figure 5.2.1a.

The maximal horizontal shear forces N_{xy} at the shell corners can be calculated using Eq. (5.2.16). The diagram of these forces is shown in Figure 5.2.1b:

$$N_{xy} = 2.04\, F_d R_y a/b = 2.04 \cdot 8.8 \cdot 4.35 \cdot 12/3 = 312.4 \text{ kN/m}$$

The main tension forces N_1 and their direction are calculated according to Eqs. (5.1.2) ... (5.1.4). The forces are maximal at the shell corner and their values are equal to the maximal horizontal shear forces N_{xy}:

$$N_{m1} = -N_{m2} = 312.4 \text{ kN/m}$$

$$-\alpha_1 = \alpha_2 = 45°$$

The maximum transverse bending moment, M_y, along the shell vertex is calculated according to Eq. (5.2.19). The moments' diagram is presented in Figure 5.2.1c.

$$\max M_y = -0.2\, F_d\, h^2\, R_y^2/b^2 = -0.2 \cdot 8.8 \cdot 0.1^2 \cdot 4.35^2/3^2 = -0.037 \text{ kN m/m}$$

The local bending moment, M_x, that appears near the diaphragms is calculated in a same way like for translation shells (see Section 4.2.5 and Figure 4.2.9a in Chapter 4). The useful expressions are Eqs. (4.2.40) ... (4.2.45). The maximal negative bending moment (Figure 4.2.9e) is:

$$M_x = -0.289\, F_d R_y\, h = -0.289 \cdot 8.8 \cdot 4.35 \cdot 0.1 = -1.11 \text{ kN m/m}$$

This moment becomes equal to zero at distance c_x from the diaphragm:

$$c_x = 0.60 \sqrt{R_y\, h} = 0.6\,(4.35 \cdot 0.1)^{1/2} = 0.4 \text{ m}$$

5.4.3 Calculating the buckling load for a covering shell

Structural aspects of the covering shell of the 2E terminal at Charles de Gaulle Airport in Paris were described in Section 5.1.3. The design load for this shell, including snow, is 13 kN/m² (the part of the self-weight, dead and service loads is about 12 kN/m²). As it was mentioned in Section 5.1.3, the shell was in use more than two years, i.e., the influence of long-term loading should be considered. Following the general theory of shells (Billington 1990; Fisher 1968), this shell is thin-walled, as the ratio between its thickness and short span is $h/l_2 \leq 1/100$. In such shells, the long-term loading influence is very important due to their geometrical and physical non-linearity (Iskhakov and Ribakov 2014; Tomas and Tovar 2011; Shugaev and Sokolov 2011). The shell buckled on 23 of May 2004 in the middle of its long span. The buckling zone length is about 30 m and its width is equal to that of the shell (Figure 5.4.1).

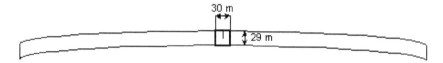

Figure 5.4.1. The buckling zone of the shell (see also Figure 5.1.3).

In order to select the appropriate design case, let us check the parameter *m* according to Eq. (5.2.23):

$$l_2 = 26 \text{ m}, \quad b = 13 \text{ m}, \quad R_y = 24.7 \text{ m}, \quad s_y = 2\,\alpha_0\,R_y = 2 \cdot 0.554 \cdot 24.7 = 27.4 \text{ m}$$

$$\alpha_0 = \text{arc sin}\,(b/R_y) = \text{arc sin}\,(13/24.7) = 31.8° = 0.554 \text{ rad}$$

$$m = 1.76\alpha_0\,\sqrt[4]{\frac{R_y^3}{l_2^3}} = 1.76 \cdot 0.554\,\sqrt[4]{\frac{24.7^3}{26^3}} = 0.938 < 1$$

As $m \leq 1$, the critical load for the shell is calculated according to Eq. (5.2.24):

$$q_{cr} = E_c\,(t)\,\frac{h}{R_y}\left[0.857\,\frac{h^2}{s_y^2} + 0.101\,\frac{s_y^2\,h^4}{R_y^2\,l_2^4}\right] =$$

$$= E_{c.red}\,\frac{0.3}{24.7}\left[0.857\,\frac{0.3^2}{27.4^2} + 0.101\,\frac{27.4^2 \cdot 0.3^4}{24.7^2 \cdot 26^4}\right] = 0.014\,10^{-4}\,E_{c.red}$$

The elasticity modulus for the shell concrete, E_c, reduces in time due to concrete creep and for the discussed case it is

$$E_{c.red} = 0.319\,E_c$$

Assuming a zone with humidity higher than 40% and that the shell is made of concrete class C 30, follows:

$$E_{c.red} = 0.319\,E_c = 0.319 \cdot 27000 = 8613 \text{ MPa} = 8613\,10^3 \text{ kN/m}^2$$

$$q_{cr} = 0.014 \cdot 10^{-4}\,E_c\,(t) = 0.014 \cdot 10^{-4} \cdot 8613 \cdot 10^3 = 12 \text{ kN/m}^2$$

If there is no snow load (the shell had collapsed on May 23), the design load for the shell is about 12 kN/m^2. Some additional issues should be taken into account:

- the shell has light openings that significantly reduce its section;
- the edge element stiffness in horizontal direction is not enough to carry the thrust forces, transferred from the shell;
- there is an additional self-weight, contributed by the external steel pipes over the shell.

Each of these issues individually or all together could be a reason for the shell buckling.

6

Hyperbolic Paraboloid Shells

6.1 Saddle shells

6.1.1 Saddle surface equation

A saddle surface is a translation one that includes two systems of arches—
one convex (direction x in Figure 6.1.1) and the other concave (direction y).

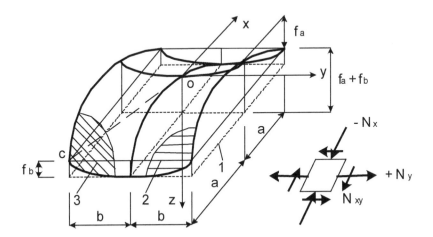

Figure 6.1.1. A saddle shaped shell: a system of internal forces and reinforcement bars according to calculation: 1—basic rectangle; 2—reinforcement bars, A_{sy}; 3—diagonal reinforcement bars.

The surface equation is:

$$z = f_a \frac{x^2}{a^2} - f_b \frac{y^2}{b^2} \qquad (6.1.1)$$

For example, the rise of point c at the shell corner (where $x = -a$, $y = -b$) relative to point o is:

$$z_c = f_a - f_b \qquad (6.1.2)$$

The lines that connect the surface center, o, and the corners (for example co, shown in Figure 6.1.1) are straight ones, i.e., the surface is a linear one in diagonals' directions (hyperbolic paraboloid; see also Section 1.2.4).

6.1.2 Calculation of the shell

As a saddle surface is a translation one and includes two systems of parabolas: convex and concave (see Figure 6.1.1). It is possible to calculate such shells according to the methodology explained in Chapter 4. The main normal and torsion curves for such shells are:

$$k_1 = k_x = \frac{\partial^2 z}{\partial x^2} = \frac{2 f_a}{a^2}; \; k_2 = k_y = \frac{\partial^2 z}{\partial y^2} = -\frac{2 f_b}{b^2}; \; k_{xy} = \frac{\partial^2 z}{\partial x \, \partial y} = 0 \quad (6.1.3)$$

As the curves in y direction are negative (concave parabolas), the membrane forces in this direction, N_y, are tensile ones, therefore reinforcing bars by calculation should be used (see lines 2 in Figure 6.1.1):

$$A_{sy} = N_y / f_{sd} \qquad (6.1.4)$$

The membrane forces, N_x, in the positive curves' direction (convex parabolas) are compression ones, like in shells with positive Gaussian curvature, therefore the reinforcement bars in this direction are according to constructive requirements (EN 1992). The main forces are calculated, similar to translation shells, and diagonal reinforcing bars are provided (see lines 3 in Figure 6.1.1).

6.2 Hyperbolic shells

6.2.1 Simple hyperboloid

The easiest way to obtain a simple hyperboloid is:

- to draw a rectangle on a horizontal plane;
- to rise one of the rectangle's corners together with the sides that are connected to it above the plane, keeping the sides straight;

- the surface that is obtained by line translation, parallel to one of the sides along two perpendicular edges (directional lines), is a simple hyperboloid similar to a propeller;
- according to this definition, the surface of a simple hyperboloid is linear (Figure 6.2.1a).

As simple hyperboloids have linear surfaces, they are very popular, because it is possible to carry out the construction works using planar wood plates. In spite of the fact that these surfaces are created by straight lines, they have negative double curvature (K < 0) and are non-developable. Horizontal sections of such surfaces are hyperbolic lines and their asymptotes are the sides of the basic rectangle a and b (see Figure 6.2.1).

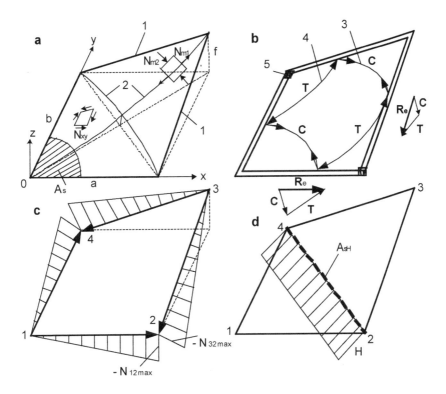

Figure 6.2.1. A simple hyperboloid:

a—geometry of the hyperboloid; b—reactions in the edge elements; c—compression forces diagram in edge elements (similar to axial forces in columns due to self-weight); d—thrust forces, H, in the direction of diagonal 2–1 and reinforcement steel A_{sH}; 1—straight lines; 2—convex and concave parabolic lines; 3—compression arches; 4—cables in tension; 5—columns at corners 2 and 4.

6.2.2 Calculation of a simple hyperboloid

The surface equation of a simple hyperboloid (see Figure 6.2.1a) is:

$$z = \frac{f}{a\,b}\,x\,y \qquad (6.2.1)$$

The curves of the linear hyperboloid shell are:

$$k_x = \frac{\partial^2 z}{\partial x^2} = 0;\; k_y = \frac{\partial^2 z}{\partial y^2} = 0;\; k_{xy} = \frac{\partial^2 z}{\partial x\,\partial y} = \frac{f}{a\,b} = \text{const} \qquad (6.2.2)$$

i.e., the shell has a pure constant torsion curvature as a propeller.

In this case, the system of differential equations (3.2.12) for a group of membrane forces takes a simple form ($N_{xy} = -\,d^2\varphi/dx\,dy,\, k_{xy} = f/(a\,b)$):

$$\frac{2\,f}{a\,b}\frac{\partial^2\varphi}{\partial x\,\partial y} = q \qquad (6.2.3)$$

where $\varphi\,(x, y)$ is the stresses' function;

q is a uniformly distributed load, acting on the shell.

Let's select the function in the following form:

$$\varphi\,(x, y) = q\,a\,b\,x\,y/(2\,f) \qquad (6.2.4)$$

This expression corresponds to the differential equation (6.2.3) and to the following boundary conditions:

when $x = \pm\,a$

$$N_x = \frac{\partial^2\varphi}{\partial y^2} = 0 \qquad (6.2.5)$$

for $y = \pm\,b$

$$N_y = \frac{\partial^2\varphi}{\partial x^2} = 0 \qquad (6.2.6)$$

The above conditions show that in corresponding directions no stiffness of the shell edge elements is required. At the same time, the normal membrane forces, N_x and N_y, are equal to zero at the entire surface of the shell, which follows from the two first expressions (6.2.2). The shear membrane forces, N_{xy} (horizontal distortion in the differential element's plane—see Figure 6.2.1a), are calculated as:

$$N_{xy} = \frac{\partial^2\varphi}{\partial x\,\partial y} = \frac{q\,a\,b}{2\,f} = \text{const} \qquad (6.2.7)$$

i.e., constant shear forces correspond to constant torsion curvature of the shell (pure shear). Therefore, the main forces are also equal to N_{xy}:

$$N_{m1} = -N_{m2} = \frac{q\,a\,b}{2\,f} = N_{xy} = \text{const} \tag{6.2.8}$$

The directions of these forces are:

$$-\tan 2\alpha_1 = \tan 2\alpha_2 = \frac{2\,N_{xy}}{N_x - N_y} = \infty; \, -\alpha_1 = \alpha_2 = 45° \tag{6.2.8a}$$

The reinforcement bars, corresponding to the main tensile forces, N_{m1}, are provided along the entire surface of the shell in parallel with the concave parabolas (see Figure 6.2.1a):

$$A_s = N_{m1}/f_{sd} \tag{6.2.9}$$

The shear forces, N_{xy}, along the shell's perimeter are transferred to the edge elements—see reactions R_e in Figure 6.2.1b. These elements are subjected to pure axial compression due to self-weight (Figure 6.2.1c). They are calculated, according to the following expressions:

$$N_{12\,max} = -N_{xy}\,a; \quad N_{14\,max} = -N_{xy}\,b \tag{6.2.10}$$

$$N_{32\,max} = -N_{xy}\,\sqrt{b^2 + f^2}; \quad N_{34\,max} = -N_{xy}\,\sqrt{a^2 + f^2} \tag{6.2.11}$$

Thrust forces, H, appear in the direction of diagonal 2–4. These forces (Figure 6.2.1d) are calculated as:

$$H = \sqrt{N_{12\,max}^2 + N_{32\,max}^2} \tag{6.2.12}$$

The corresponding reinforcement steel section is:

$$A_{sH} = H/f_{sd} \tag{6.2.13}$$

6.2.3 Composite hyperboloid

An effective way to use hypars is a system of four simple hypars, called composite hypar. Such hypar is usually symmetric (Figure 6.2.2a–e) and supported on a symmetric system of columns. In a composite hypar it is possible to add four tensile bars in order to take the thrust forces (Figure 6.2.2f for the composite hypar, corresponding to scheme d). Alternatively, two intersecting bars can be provided. If there is no possibility to use tensile bars, the thrust forces should be taken by the frame itself.

To cover big areas it is possible to combine composite hypars (consisting of four simple ones). For example, the dimensions of the roofing at the Basis for Absorption and Classification in Tel-Hashomer, Israel, are 18 × 24 m.

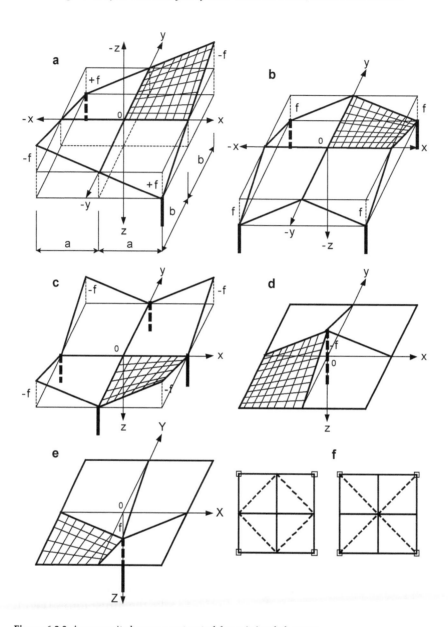

Figure 6.2.2. A composite hypar, constructed from 4 simple hypars:

In cases a, b, c the central point, o, remains at the hypars' intersection, and four corners are risen or lowered; in cases d and e the central point is risen or lowered, but the corners remain at the original height; f—diagonal bars (dashed lines) for taking the thrust forces in case d.

The roofing is constructed using six composite hypars 8 × 9 m each in a form of an inverse umbrella (see Figure 6.2.2e). A scheme of the structure is presented in Figure 6.2.3. The composite hypar consists of four simple ones—two of them are 3 × 4 m and two others—4 × 6 m. The thickness

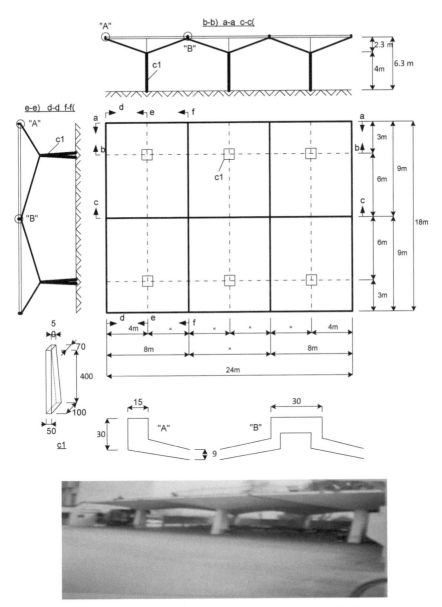

Figure 6.2.3. Roofing of the Absorption and Classification Center in Tel-Hashomer, Israel, made of 6 composite hypars, i.e., 24 simple ones.

of each hypar is 9 cm. The roofing is supported on six columns in a form of a truncated pyramid—the net of columns' axes is 8 × 12 m. Every two adjacent simple hypars have a common edge element. The structure is made of concrete class C 30 with a density of 22.8 kN/m³.

6.2.4 General approaches for calculating composite hypars

Calculation of a composite hypar depends on its constructive scheme (see Figure 6.2.2). For example, for a special case, shown in Figure 6.2.2d, the equation of the rectangular shell surface is:

$$z = -f_0 \left(1 - \frac{x+y}{a} + \frac{x\,y}{a^2} \right)$$ (6.2.14)

where a is the span of the simple rectangular hypar;

f_0 is the rise of the composite hypar (equal to that of the simple one).

The equation is also suitable for to an "inverse umbrella" surface (see Figure 6.2.2e), but without a minus in the beginning (see Eq. 6.2.14).

In any case, calculation of a composite hypar begins from the simple ones, according to Section 6.2.2. If a composite hypar corresponds to schemes d or e in Figure 6.2.2, i.e., the edge elements along the shell perimeter are horizontal, then the walls' upper line will be horizontal, which is convenient from the construction point of view. In this case, there is no need for additional calculation of the composite hypar. It should be taken into account that the common edge elements between two adjacent simple hypars take double pure tension or compression, compared with a simple one.

Triangular trusses can be provided, as diaphragms along the perimeter of a composite or simple hypar (see Figure 6.2.2a). Each of them has a length of two simple hypars, as shown in Figure 2.1.2d. Alternatively, walls that carry the horizontal shear forces, transferred from the hypar, can be used (Figure 6.2.4).

Let's assume that a composite hypar is supported by three columns at corners A, B and D (the column at corner A is used just for obtaining the system's stability) and four trusses along the perimeter (Figure 6.2.5). If the external load is symmetric relative to axis BD then $R_A = 0$.

The horizontal shear forces, N_{xy}, transferred from the shell to the truss, cause stresses only in its upper belt; there are no stresses in the bottom belt and the diagonal bars. The upper belt of the truss is in compression. It is clear that:

$$\Sigma M_B = 0; R_A \, 2\,a = 0; R_A = 0$$

Figure 6.2.4. Auditorium at the Hadassa-Ein-Cerem Hospital in Jerusalem.

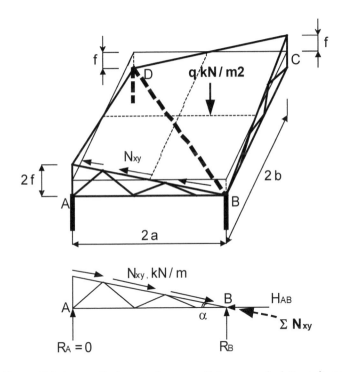

Figure 6.2.5. A truss diaphragm of a composite hypar—calculation scheme.

The horizontal and the vertical reactions at support B are:

$$H_{AB} = \Sigma N_{xy} \cos \alpha; \quad R_B = \Sigma N_{xy} \sin \alpha \qquad (6.2.15)$$

where

$$\Sigma N_{xy} = N_{xy} \, 2 \, a / \cos \alpha \qquad (6.2.16)$$

If there are no diaphragms along the perimeter of the shell, a tensile bar should be used to take the thrust force. This force in direction BD (convex direction) is:

$$H_{BD} = (H_{AB}^2 + H_{BC}^2)^{1/2} \qquad (6.2.17)$$

The section of the tensile bar (dashed line in Figure 6.2.5) is:

$$A_{sH} = H_{BD}/f_{sd} \qquad (6.2.18)$$

6.3 Numerical examples for hypars

6.3.1 Simple hypar—calculating the internal forces and the reinforcement section

Given a simple RC hypar, as shown in Figure 6.2.1. The hypar has a square basis $a = 9$ m, its rise $f = 1.8$ m and thickness $h = 10$ cm. There is a tensile bar between corners 2 and 4 for taking the shell thrust forces (see Figure 6.2.1d). The shell is made of concrete class C 30 and ribbed steel bars. The shell carries a dead load of 0.5 kN/m² and live load of 0.5 kN/m² additionally to its dead weight.

It is required to calculate the hypar.

Solution

The dead weight of the shell is:

$$g_k = 0.1 \cdot 24 = 2.4 \text{ kN/m}^2$$

The design load is:

$$q = F_d = 1.4 \, (g_k + \Delta g_k) + 1.6 \, q_k = 1.4 \, (2.4 + 0.5) + 1.6 \cdot 0.5 = 4.9 \text{ kN/m}^2$$

The internal forces are calculated according to Eqs. (6.2.7 ... 6.2.13).

$$N_{xy} = \frac{q \, a \, b}{2 \, f} = 4.9 \cdot 9^2/(2 \cdot 1.8) = 110.25 \text{ kN/m} = \text{const}$$

$$N_{m1} = -N_{m2} = N_{xy} = 110.25 \text{ kN/m} = \text{const}$$

$$-\alpha_1 = \alpha_2 = 45°$$

The reinforcement section in the main tension forces' direction, acting along the concave diagonal (see Figure 6.2.1a), is:

$$A_s = N_{m1}/f_{sd} = 110.25 \cdot 10^3/350 = 315 \text{ mm}^2/\text{m} = 3.15 \text{ cm}^2/\text{m}$$

The problem is that the above-mentioned reinforcement bars have different lengths. It is also possible to use straight reinforcement bars in x and y directions in a form of a net (in this case, the bars' length is constant). In this case the reinforcement sections are:

$$A_{sx} = A_{sy} = A_s/2^{1/2} = 3.15/1.41 = 2.23 \text{ cm}^2/\text{m}$$

The horizontal shear forces along the perimeter of the shell, N_{xy}, act on diaphragms (see Figure 6.2.1b), which are in pure compression (Figure 6.2.1c).

$$N_{12 \, max} = N_{14 \, max} = -N_{xy} a = -110.25 \cdot 9 = -992.25 \text{ kN}$$

$$N_{32 \, max} = N_{34 \, max} = -N_{xy} \sqrt{a^2 + f^2} = -110.25 \, (9^2 + 1.8^2)^{1/2} = -1011.9 \text{ kN}$$

Thrust forces, H, appear in the direction of diagonal 2–4 (Figure 6.2.1d):

$$H = \sqrt{N_{12 \, max}^2 + N_{14 \, max}^2} = 992.25 \cdot 2^{1/2} = 1403.25 \text{ kN}$$

The reinforcement section required to take these forces, is:

$$A_{sH} = H/f_{sd} = 1403.25 \cdot 10^3/350 = 4009.3 \text{ mm}^2 = 40.1 \text{ cm}^2$$

6.3.2 Composite hypar—calculating the edge element and the thrust force

Given a complex hypar, as shown in Figure 6.2.5. The hypar consists of four simple ones and has the following dimensions: $2 a \times 2 b = 18 \times 18$ m, its rise is $2 f = 3.6$ m (where f is the rise of a single hypar).

It is required to calculate the shell diaphragm according to the static scheme shown in Figure 6.2.5.

Solution

The calculations are performed according to Eqs. (6.2.15) ... (6.2.18).

$$\tan \alpha = 2 f/(2 a) = 2 \cdot 1.8/(2 \, 9) = 0.2; \, \alpha = 11.3°$$

Following Eq. (6.2.16), the compression force in the diaphragm (in the upper belt of the truss) is:

$$\Sigma N_{xy} = N_{xy} \, 2 \, a/\cos \alpha = 110.25 \cdot 2 \, 9/\cos 11.3° = 2023.73 \text{ kN}$$

The horizontal and vertical reactions in the diaphragms are:

$$H_{AB} = \Sigma N_{xy} \cos \alpha = 2023.73 \cos 11.3° = 1984.5 \text{ kN}$$
$$R_{B} = \Sigma N_{xy} \sin \alpha = 2023.73 \sin 11.3° = 396.54 \text{ kN}$$

The thrust force in the convex direction of the shell, BD, is:

$$H_{BD} = (H_{AB}^{2} + H_{BC}^{2})^{1/2} = 2^{1/2} H_{AB} = 2^{1/2} \cdot 1984.5 = 2806.5 \text{ kN}$$

The required section of the tensile steel (dashed line in Figure 6.2.5) is:

$$A_{sH} = H_{BD}/f_{sd} = 2806.5 \cdot 10^{3}/235 = 11943 \text{ mm}^{2} = 119.43 \text{ cm}^{2}$$

7

Shells of Revolution— Domes

7.1 World-famous dome structures

7.1.1 Millennium dome in London

The Millennium Dome is constructed in the Greenwich district of London near Thames river. The structure was a complex and complicated building operation—it is the biggest dome in the world. The construction process started in 1997 and completed in 1999. The diameter of the shell is 320 m and its rise above the ground is 50 m (Figure 7.1.1). The shall perimeter is 1005 m, the dome covers an area of 80425 m². The radius of curvature of the dome is 281 m and a half of its central angle is 34.7°.

The dome that costs 750 million Pounds was constructed with the aim to store the huge multimedia exhibition of British technology along its whole history. Now the structure has become a new center of ceremonies and business in London.

The dome is pre-cast (the pre-cast elements are made of transparent material) and it is suspended on 12 masts that have a 100 m height and the shape of each one is similar to banana. Every mast takes the forces, transferred by 120 cables, supporting one strip, i.e., each cable carries the self-weight of a strip that covers an area of 56 m². The total length of all cables is more than 72 km. The dome is composed of PTEF strips that are two-layer glass (fiberglass). Two layers of the strips allow preventing condensation.

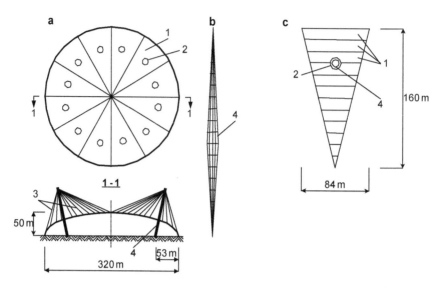

Figure 7.1.1. The Millennium Dome: a—upper view; b—banana-shaped mast; c—strips, consisting of 12 parts; 1—shell; 2—openings for masts; 3—cables; 4—one of the 12 masts.

7.1.2 Aqua Park dome in Moscow

The covering of the Aqua Park in Moscow (Figure 7.1.2a) is a revolution shell—part of a dome (a little bit more than a quarter). The line that forms the shell surface consists of two curves, 1 and 2 that have a common tangent at their connection point A. The rise of the dome is 11 m. Lines 1 and 2 have circular arch shapes with radiuses of 163.8 m and 277.8 m, respectively (see Figure 7.1.2e), which was the main idea of the designers. There are two kinds of surfaces in the shell—with double and single curvatures (lines 1 and 2, correspondingly).

The upper view of the shell' geometry is shown in Figure 7.1.2b and the vertical cross section is shown in Figure 7.1.2c. The shell is made of reinforced concrete (concrete class C 35, modulus of elasticity 34500 MPa) and has a 7 cm thickness in the span and 20 cm one near the footing ring. The structure has a system of ring ribs (each 6 m) and meridian ones of 33 cm height and 12 cm width (25 cm near supports; Figure 7.1.2d). The mass of the RC dome is 2500 t and its area is more than 4000 m², the self-weight is about 6 kN/m². The average dome thickness is 25 cm.

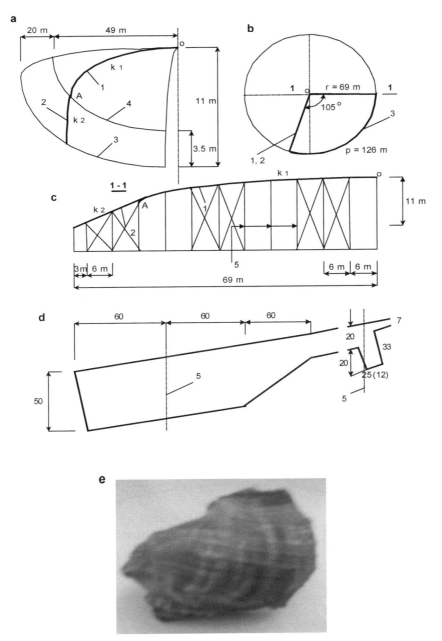

Figure 7.1.2. Geometry and structural scheme of the Aqua Park covering dome in Moscow: a—a 3D view; b—upper view; c—side view; d—supporting ring and ribs sections; e—natural shell; 1—curved line k_1; 2—curved line k_2; 3—supporting ring; 4—border between the surfaces; 5—column axis.

The shell is supported on a ring with a section that is 50 cm in height and 120 cm in width. The ring is supported on columns made of Ø426 mm and 9 mm thickness steel pipes. The connections between the ring and the columns as well as between the columns and the foundations are hinged immovable. There are X-shaped braces between the columns. The steel bars in the supporting ring are 36Φ36 and their characteristic strength is 500 MPa. The reinforcement net in the dome is made of Φ12 steel bars and Φ25 ones in the ribs.

7.1.3 Geodetic dome of the Ice Park in Eilat

A geodetic surface is close to spherical one and consists of equiangular triangular, *pentagon* or *hexagon* planar parts. The sides of the adjacent parts are coincided with the dome. This geodetic shell is composed of triangular elements, made of wood, forming a structure, in which the ratio between the material quantity and the volume is minimal. Geodetic shells have practically no limit in dimensions.

By using a big number of basic units, it is possible to construct shells with very big radius, so that they won't buckle under their self-weight. Various construction materials, including natural ones, like wood, bamboo, etc. are used for building geodetic shells.

The diameter of the dome in Eilat is 106 m (Figure 7.1.3). According to the design, 132 triangles were required to construct the dome. The side length of each triangle is 12.5 m (Figure 7.1.4).

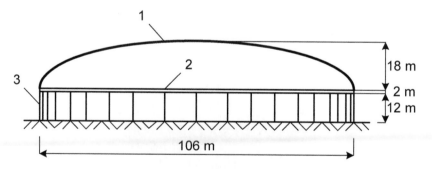

Figure 7.1.3. The geodetic dome in Eilat: 1—the dome made of wood; 2—RC supporting ring; 3—columns.

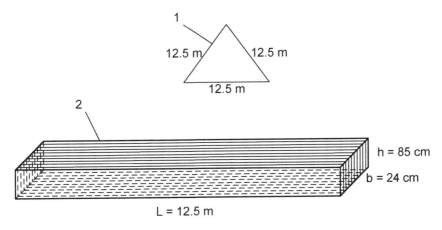

Note: In order to show the layers the beam width is drawn out of scale.

Figure 7.1.4. The triangle side: 1—a basic triangle; 2—multi-layer wood beams.

The weight of the multi-layer wood beam for the given span and rectangular section dimensions (24 × 85 cm) is about 2 t. Multi-layer wood beams are composed of a number of laminated beams, controlled before gluing to avoid "eyes" or cracks. Figure 7.1.5 presents a 3D model of the dome and a detail of steel connectors in the bottom part that unify the edges of six beams of the adjacent six triangles.

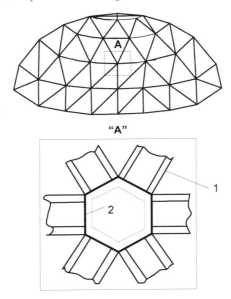

Figure 7.1.5. A 3-D model of the dome and a connection detail (2), unifying edges of the triangles' beams (1).

The wood shell has a rise $f = 18$ m, i.e., it is a shallow shell: $f/D \approx 1/6$. The dome is supported on RC ring with a rectangular section that has a height of 2 m and width of 1 m. The construction cost of this dome was about 5 million US dollars. 12% of the shell surface area is made of transparent skylight triangles. These triangles include automatic openings that enable smoke offtake from the building (Figure 7.1.6).

The dome is covered by a perforated thermo-acoustic sandwich. Above the sandwich is an acoustic isolation, above which is a thermic isolation, covered by OSP plates made of laminated wood layers. Above all these layers is a water isolation and above it is a covering layer made of a waved zinc coated steel sheet (Figure 7.1.7).

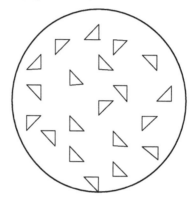

Figure 7.1.6. 12% transparence in the dome area.

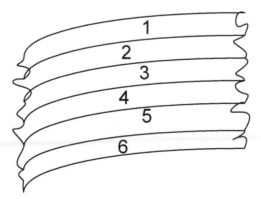

Figure 7.1.7. Thermo-acoustic-water protection of the shell roof (non-scaled):

1—waved sheet covering; 2—water isolation; 3—covering by OSP plates; 4—thermic isolation; 5—acoustic isolation; 6—wood shell.

The dome is designed to carry its self-weight and earthquake loads, because Eilat is located in a zone with rather high seismic hazard. It is also designed to withstand wind loads, because there are strong winds at the Eilat gulf. The structure should also carry the load of 90 kN from the artificial lighting equipment, installed at different locations.

The main wood triangle includes secondary wood triangle and filling wood plates (Figure 7.1.8).

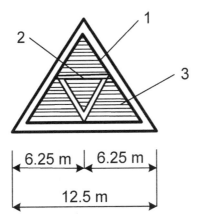

6.25 m | 6.25 m

12.5 m

Figure 7.1.8. Main (1) and secondary (2) triangles and filling wood plates (3).

The views of the dome in Eilat are given in Figure 7.1.9.

Let's calculate the average thickness of the dome. The wood volume in one triangle (main, secondary and filling):

$$v_{wm} = 3 \cdot 12.5 \cdot 0.24 \cdot 0.85 + 3 \cdot 6.25 \cdot 0.12 \cdot 0.425 + 0.05 \cdot 0.5 \cdot 12.5 \, (12.5 \sin 60°) =$$

$$7.65 + 0.96 + 3.38 = 11.99 \text{ m}^3$$

The main triangle area is:

$$S = 0.5 \cdot 12.5 \, (12.5 \sin 60°) = 67.66 \text{ m}^2$$

The average dome thickness is:

$$h_{wm} = 11.99/67.66 = 0.18 \text{ m}$$

The radius of curvature at the vertical section is:

$$R = [(D/2)^2 + f^2]/(2f) = [(106/2)^2 + 18^2]/(2 \cdot 18) = 87 \text{ m}$$

If modulus of elasticity for wood is E_w = 10,000 MPa, then the lower limit of the buckling load for this shell is

$$q_{cr} = 0.2 \, E_w \, (h_{wm}/R)^2 = 0.2 \cdot 10.000 \, (0.18/87)^2 = 0.00856 \text{ MPa} = 8.56 \text{ kN/m}^2.$$

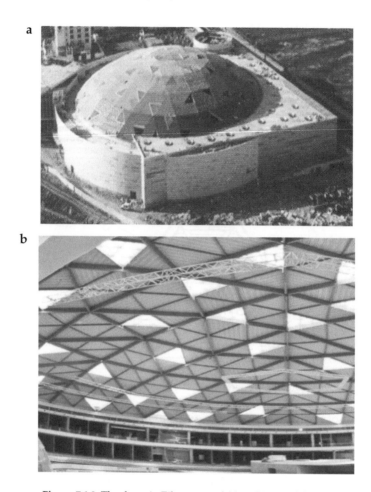

Figure 7.1.9. The dome in Eilat: external (a) and internal (b) views.

7.1.4 Reinforced concrete elliptic-shape shell in Switzerland

An elliptic RC shell of revolution was constructed in Chiasso, Switzerland in 2012. The main dimensions of the shell are $22 \times 52 \times 93$ m (Figure 7.1.10). The shell thickness varies from 10 to 12 cm. The structure is made of casted and sprayed concrete. It was constructed using modern steel fibered concrete technologies and recently developed methods for pre-stressing thin-walled curved structures, as well as numerical formworks' design approaches. The shell reinforcement includes four layers of steel nets (two in the upper and two in the bottom section zone). The directions of reinforcement bars in the nets correspond to those of the main ellipsoid axes—ribs and rings.

There is a compressed supporting ring in the basis of the shell. The ring is supported on inclined columns (tangent to the shell ribs).

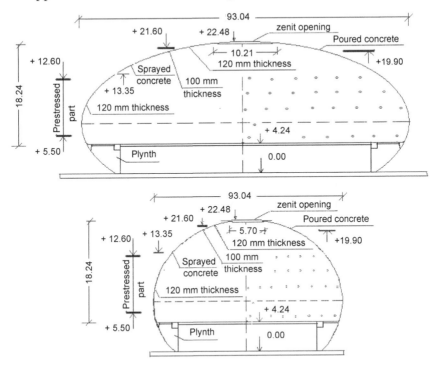

Figure 7.1.10. Main dimensions of the shell: a—in longitudinal direction; b—in transverse direction (following Muttoni et al. 2012).

In addition to ordinary steel bars, there are 35 pre-stressed cables (5 cables × 7 parts of line). Each cable has a diameter of 0.6 inch and consists of 19 wires. The cables were installed in the equator zone of the shell, where the membrane tensile forces appear in the horizontal ring. The height of this zone is between 5.5 m and 12.6 m from the floor of the structure (not from the shell basis). The shell thickness in this zone is increased from 10 cm to 12 cm to provide proper covering of the pre-stressing cables by concrete. The concrete class is C 30 for all elements of the shell. The structure has an upper opening with dimensions of 5.7 × 10.21 m and many circular openings with a diameter of 0.4 m.

The ratio of hooked steel fibers in the sprayed concrete was 30 kg/m³. The fibres' length is 30 mm and their length to diameter ratio is 80. Fibers were used in order to improve the concrete cracking resistance (first of all in the pre-stressing zone) and to increase its ductility under the membrane forces in the shell, as well as local bending moments and shear forces in the connection zone between the shell and the supporting ring.

One of important problems in the design of the shell was the buckling of its upper zone that has the lowest curvature in both main directions. The membrane forces in this zone get their maximum values in compression. The shell was calculated, considering physical concrete non-linearity (creep), as well as geometrical non-linearity of the thin-walled shell (decrease in the Gaussian curvature of the surface).

The construction works were carried out using timber formwork, placed over timber false-work. After completing the formwork assembling, reinforcement nets and cables were placed and concrete was casted and sprayed. The duration of these processes was about three months. The overall construction cost is subdivided as follows:

- 49% for the formwork;
- 21% for ordinary reinforcement;
- 5% for post-pre-stressing;
- 24% for sprayed concrete;
- 1% for casted concrete.

In other words, the part of the formworks' cost is very high. Therefore, alternative methods with more reasonable price should be developed.

7.2 Statically determined spherical shells

7.2.1 General

A spherical dome is a part of a sphere and its geometry is the most popular one in shells of revolution. From the viewpoint of Gaussian geometry, the dome surface has a positive curvature (K > 0), i.e., the surface has a double positive curvature. Generally, the system of coordinates of a dome surface includes meridians and parallels, similar to those of the Earth. The parallels (rings) are horizontal (parallel to the covered area).

There are two constructive requirements to thin-walled shallow domes:

$$1/300 > h/D > 1/500 \tag{7.2.1}$$

$$f/D \leq 1/10 \tag{7.2.2}$$

where D is the diameter of the dome;

h is its thickness;
f is the dome height above the footing ring.

If the dome is simply supported by hinge movable supports (at any point of the supporting ring) and the direction of these supports corresponds to that of the tangents to the meridian at the same point, the dome is statically determined for axially symmetric load. It means that it is enough to use just the equilibrium equations, in order to calculate the supports' reactions and the internal forces in the dome.

In thin-walled domes subjected to uniformly distributed load, internal membrane forces N_1 (in the meridian direction) and N_2 (in the rings' direction) appear—Figure 7.2.1a. Significant bending moments appear in a narrow belt near the supporting ring (local moments).

7.2.2 Calculation of membrane forces

Let's calculate the membrane forces at any point of the dome determined by an angle coordinate φ (Figure 7.2.1b), so that

$$0 \le \varphi \le \varphi_0 \tag{7.2.3}$$

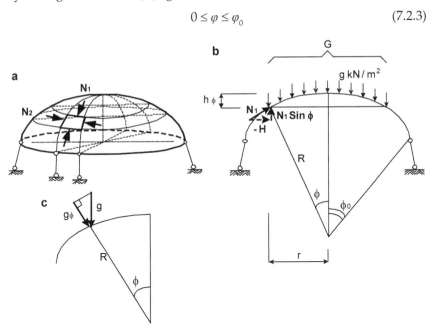

Figure 7.2.1. Analysis of statically determined dome:

a—membrane forces; b—static scheme; c—the radial loading influence.

Uniformly distributed dead load (including self-weight)

At a certain point of the dome, corresponding to coordinate φ, there is an internal membrane force, N_1, along the entire ring with a radius r

$$\Sigma N_1 = 2\pi r N_1 \tag{7.2.4}$$

The vertical component of this force is:

$$\Sigma N_1 \sin \phi = 2\pi r N_1 \sin \phi = 2\pi R \sin \phi N_1 \sin \phi = 2\pi R N_1 \sin^2 \phi \tag{7.2.5}$$

At the same time, the external load that acts on the dome over the ring with a radius r (segment with radius r) is:

$$G = S g = 2\pi R h_\phi g = 2\pi R R (1 - \cos \phi) g = 2\pi R^2 (1 - \cos \phi) g \tag{7.2.6}$$

where S is the area of the circular segment of radius r and rise h_ϕ.

From the equilibrium of the external and internal forces yields:

$$\Sigma N_1 \sin \phi = G \tag{7.2.7}$$

Therefore,

$$N_1 = g R \frac{1 - \cos \varphi}{\sin^2 \varphi} \tag{7.2.8}$$

Let's simplify the trigonometric expression in this equation:

$$\frac{1 - \cos \varphi}{\sin^2 \varphi} = \frac{2 \sin^2 (\varphi/2)}{\sin^2 \varphi} = \frac{1}{2 \cos^2 (\varphi/2)} = \frac{1}{1 + \cos \varphi} \tag{7.2.9}$$

Taking into account that

$$\sin^2 \phi = (2 \sin \phi/2 \cos \phi/2)^2; \quad \cos^2 \phi/2 = \{[(1 + \cos \phi)/2]^{0.5}\}^2 = (1 + \cos \phi)/2$$

and considering that N_1 is a compression force,

$$N_1 = -\frac{g R}{1 + \cos \varphi} \tag{7.2.10}$$

The horizontal component of this force with an opposite sign equals to the shell trust force at the given point:

$$H = -N_1 \cos \varphi = g R \frac{\cos \varphi}{1 + \cos \varphi} \tag{7.2.11}$$

To calculate the ring membrane force N_2, let's use the equation of forces' equilibrium (Eq. 3.2.13 in Chapter 3) and Figure 7.2.1c:

$$\frac{N_1}{R_1} + \frac{N_2}{R_2} = -g_\varphi \tag{7.2.12}$$

For a circular dome

$$R_1 = R_2 = R \tag{7.2.13}$$

Following Figure 7.2.1c, $g_\varphi = g \cos \varphi$

Therefore,

$$N_2 = - g\, R \left(\cos \varphi - \frac{1}{1 + \cos \varphi} \right) \qquad (7.2.14)$$

The diagrams of these forces for a half-circular dome are shown in Figure 7.2.2a.

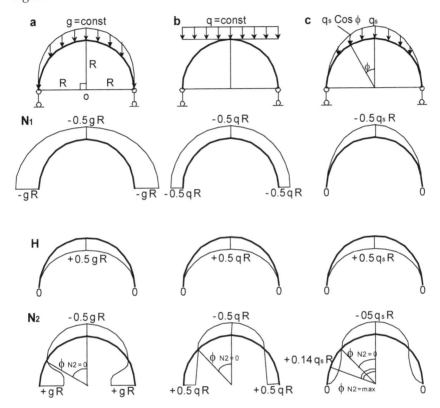

Figure 7.2.2. Internal forces' diagrams for a half-circular dome:

a—dead load; b—live load (constant on the surface horizontal projection); c—snow load.

In the diagram of N_2 is a ring with zero values of this force. This ring is determined by an angle $\varphi_{N2=0}$. To find this angle let's equate Eq. (7.2.14) to zero:

$$\cos^2 \varphi + \cos \varphi - 1 = 0 \;\; \rightarrow \;\; \cos \varphi = 0.618 \;\; \rightarrow \;\; \varphi_{N2=0} = 51.83° \quad (7.2.15)$$

Live load (constant on the surface horizontal projection)

The load is shown in Figure 7.2.2b. The calculation equations can be obtained similar to the way used for the dead load.

The calculation expressions are:

$$N_1 = -0.5\,q\,R = \text{const} \tag{7.2.16}$$

$$H = 0.5\,q\,R\cos\varphi \tag{7.2.17}$$

$$N_2 = -0.5\,q\,R\cos(2\varphi) \tag{7.2.18}$$

As $N_2 = 0$,

$$\varphi_{N2=0} = 45° \tag{7.2.19}$$

Snow load

The load is shown in Figure 7.2.2c. The calculation expressions are:

$$N_1 = -0.5\,q_s\,R\cos\varphi \tag{7.2.20}$$

$$H = 0.5\,q_s\,R\cos^2\varphi \tag{7.2.21}$$

$$N_2 = -0.5\,q_s\,R\cos(2\varphi)\cos\varphi \tag{7.2.22}$$

As $N_2 = 0$,

$$\cos(2\varphi)\cos\varphi = 0; \quad \varphi_{N2=0} = 45° \tag{7.2.23}$$

An additional problem is finding the angle $\varphi_{N2=max}$, corresponding to the maximal value of N_2:

$$d\,N_2/d\varphi = 0; \quad 2\sin(2\varphi)\cos\varphi + \cos(2\varphi)\sin\varphi = 0;$$

$$2\cdot2\sin\varphi\cos\varphi\cos\varphi + \cos(2\varphi)\sin\varphi = 0;$$

$$4\cos^2\varphi + \cos(2\varphi) = 0; \quad 4\cos^2\varphi + \cos^2\varphi - \sin^2\varphi = 0; \quad 5 - \tan^2\varphi = 0;$$

$$\varphi_{N2=max} = 65.9° \tag{7.2.24}$$

Substituting (7.2.24) into Eq. (7.2.22) yields:

$$N_{2\,max} = +0.136\,q_s\,R \tag{7.2.25}$$

The total internal forces are:

$$N_{1\,tot} = N_{1g} + N_{1q} + N_{1s}$$

$$N_{2\,tot} = N_{2g} + N_{2q} + N_{2s}$$

7.2.3 Total thrust forces in a pre-cast dome

The total thrust force for a pre-cast dome can be calculated as

$$H_{tot} = \int_{\varphi = 0}^{\varphi = \varphi_0} H \, d\varphi \qquad (7.2.26)$$

This force is finally taken by the supporting ring.

Dead load

According to Eq. (7.2.11),

$$H_{tot} = g R \int_{\varphi = 0}^{\varphi = \varphi_0} \frac{\cos \varphi}{1 + \cos \varphi} \, d\varphi = g R \left(\varphi_0 - \int_{\varphi = 0}^{\varphi = \varphi_0} \frac{d\varphi}{1 + \cos \varphi} \right)$$

$$H_{tot} = g R \left[\varphi_0 - \tan (0.5 \, \varphi_0) \right] \qquad (7.2.27)$$

For a half-spherical dome $\varphi_0 = 90°$, hence

$$H_{tot} = 0.57 \, g R \qquad (7.2.28)$$

Live load

According to Eq. (7.2.17),

$$H_{tot} = 0.5 \, q \, R \int_{\varphi = 0}^{\varphi = \varphi_0} \cos \varphi \, d\varphi = 0.5 \, q \, R \sin \varphi_0 \qquad (7.2.29)$$

For a half-spherical dome

$$H_{tot} = 0.5 \, q \, R \qquad (7.2.30)$$

Snow load

According to Eq. (7.2.21),

$$H_{tot} = 0.5 \, q_s \, R \int_{\varphi = 0}^{\varphi = \varphi_0} \cos^2 \varphi \, d\varphi = 0.25 \, q_s \, R \left[0.5 \sin (2 \, \varphi_0) + \varphi_0 \right] \qquad (7.2.31)$$

For a half-spherical dome

$$H_{tot} = 0.39 \, q_s \, R, \, kN/m \qquad (7.2.32)$$

7.2.4 Tensile force in the supporting ring

The calculation scheme of the ring is shown in Figure 7.2.3a. The ring tensile force differential is:

$$2 T = 2 \int_{0}^{\pi/2} H_{tot} \, r_0 \cos \psi \, d\psi \qquad (7.2.33)$$

where r_0 is half of the dome diameter;

dψ is a differential of the central angle in the dome horizontal projection.

$$T = H_{tot}\, r_0 \tag{7.2.34}$$

When the dome is subjected to a number of external forces, ΣH_{tot} is used in the above given equations, i.e., loads superposition principle is used (according to elastic shells' theory).

The reinforcement section in the ring is:

$$A_{sT} = T/f_{sd} \tag{7.2.35}$$

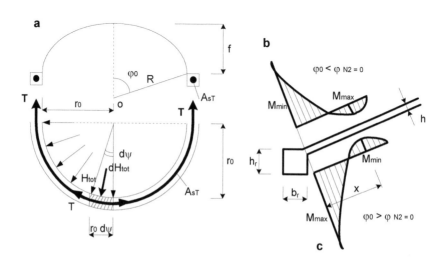

Figure 7.2.3. a—tensile force in the supporting ring; b, c—local moments between the dome and the ring.

7.2.5 Local bending moments

The local moments in the dome near the supporting ring decrease (damp) very fast (Figure 7.2.3b, c). The damping coefficient, ξ, is calculated as follows:

$$\xi = \sqrt{\frac{R}{h}}\ \sqrt[4]{3\,(1 - \mu^2)} \approx 1.3\sqrt{\frac{R}{h}} \tag{7.2.36}$$

where $\mu = 0.2$ is the Poisson's ratio.

The local moment can be obtained using an empirical expression:

$$M_{loc} = N_{20} \frac{R}{2 \, \xi^2} \tag{7.2.37}$$

where N_{20} is an algebraic value of the ring membrane force at the connection between the dome and the supporting ring (see Figure 7.2.3b).

Substituting the damping coefficient and Poisson's ratio into this equation yields:

$$M_{loc} = 0.295 \, N_{20} \, h \tag{7.2.38}$$

This moment can be positive or negative, depending on the sign of N_{20}. As it is evident from Figure 7.2.2, if

$$\varphi_0 > \varphi_{N2=0} \quad \text{then} \quad N_{20} > 0 \tag{7.2.39}$$

and the local moment is positive (Figure 7.2.3c). This case corresponds to non-shallow domes. Otherwise, if

$$\varphi_0 < \varphi_{N2=0} \quad \text{then} \quad N_{20} < 0, \tag{7.2.40}$$

and the local moment is negative (Figure 7.2.3b). This case corresponds to shallow domes.

In both cases the sign of the local moment changes and at distance x from the support and reaches its maximum value with an opposite sign:

$$x = \frac{\pi \, r_0}{2 \, \xi} \tag{7.2.41}$$

$$|M_{min}| = 0.208 \, |M_{max}| \tag{7.2.42}$$

7.2.6 Critical buckling load

Calculating the critical buckling load for domes is carried out in the same way and using the same equations, like for translation shells (see Section 4.2.4 in Chapter 4). Coefficient k in Table 4.2.1 and in Eq. (4.2.24) is equal to 1.

7.2.7 Example of a statically determined dome

The Norfolk Scope Dome (Virginia, USA) is a multi-purpose arena designed by Pier Luigi Nervi. Its construction started in 1968 and was completed in 1971. The name Scope is a shortening of "kaleidoscope" that indicates the possibility of internal changes in the structure and its adaptation to the wide variety of activities (concert hall, exhibition hall, etc.). The dome covers an area of 7900 m² (Figure 7.2.4).

Figure 7.2.4. The Norfolk Scope Dome (Virginia, USA), following (http://hamptonroads. com/2013/11/norfolk-scope-expansion-proposed-35m-cost).

The dome is a monolithic RC structure with a diameter of 134 m and rise of 15.5 m. The entire height of the building is 33.5 m. The dome is supported on 24 V-shaped columns. The columns have a slope that is tangent to the dome meridians at the supports' level (Figure 7.2.4). The supporting ring of the dome is not in tension, but behaves as a beam, transferring the meridian forces, N_{10}, to the columns. Each column takes the force from a sector that has a length of

$$\pi D/24 = \pi \, 134/24 = 17.54 \text{ m}$$

and the axial force, acting on each column, is $17.54 \, N_{10}$ kN.

7.3 A dome with elastic support along its perimeter

Let's discuss the internal forces in the connection between the dome and the supporting ring. In contrast to the structures, discussed in the previous section, generally domes are connected to supporting ring elastically. Furthermore, the supports themselves are not in the direction, tangent to the dome meridians, but vertical (Figure 7.3.1a, b). Therefore, additional forces appear along the perimeter of the dome-moment M_0 and radial thrust forces H_0, and the structure in this case will be statically undetermined.

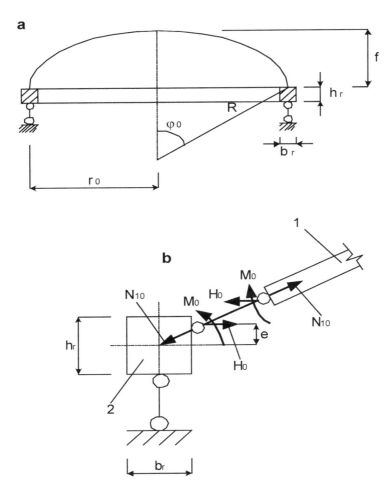

Figure 7.3.1. A static scheme of a dome (a) and its connection to the supporting ring (b): 1—the dome itself; 2—supporting ring.

These forces can be obtained from the deformations' equation of the dome and the supporting ring in radial bending and ring elongation (which is a reason for radial displacement):

$$\varepsilon_d = \varepsilon_r \qquad (7.3.1)$$

where indexes d and r denote the dome and the ring, respectively.

Defining directions of M_0 and H_0 as 1 and 2, respectively, yields the following system of flexibility method equations for calculating M_0 and H_0:

$$a_{11} M_0 + a_{12} H_0 + a_{10} = 0$$

$$a_{21} M_0 + a_{22} H_0 + a_{20} = 0 \qquad (7.3.2)$$

where a_{11} is the rotation angle in direction M_0 due to $M_0 = 1$;

a_{12} is a rotation angle in direction of M_0 due to $H_0 = 1$;

a_{21} is a displacement in direction of H_0 due to $M_0 = 1$;

a_{22} is a displacement in direction of H_0 due to $H_0 = 1$;

a_{10} and a_{20} are rotation angle and displacement of the connection due to external forces.

These coefficients can be calculated as follows (Murashev et al. 1971):

$$a_{11} = \frac{s}{D} + \frac{12\, r_0^2}{E_c\, b_r\, h_r^3}, \; \text{mm}^0/\text{N}$$

$$a_{22} = \frac{s^3}{2\,D} \sin^2 \varphi_0 + \frac{r_0^2}{E_c\, b_r\, h_r} + \frac{12\, r_0^2}{E_c\, b_r\, h_r^3}\, e^2, \text{mm}^2/\text{N}$$

$$a_{12} = a_{21} = \frac{s^2}{2\,D} \sin \varphi_0 - \frac{12\, r_0^2}{E_c\, b_r\, h_r^3}\, e, \text{mm}/\text{N} \qquad (7.3.3)$$

For an external load q (self-weight, dead load, live load)

$$a_{20} = \frac{q\, r_0\, s^4}{4\,R\,D} \left(\frac{1}{1 + \cos \varphi_0} - \cos \varphi_0 \right) \sin \varphi_0 + \frac{q\,R\,r_0^2}{E_c\, b_r\, h_r} \frac{\cos \varphi_0 \sin^2 \varphi_0}{1 + \cos \varphi_0}, \text{mm}$$

The first and second terms correspond to displacements due to the moments and the thrust forces, respectively.

$$a_{10} = \frac{q\, s^4}{2\,R\,D} \sin \varphi_0 \qquad (7.3.4)$$

This is the change in the angle, compared to the membrane state.

Following Eqs. (4.2.28 and 4.2.36),

$$s = 0.76\, (R\, h)^{1/2}; \; D = E_c\, h^3/12 \qquad (7.3.5)$$

Substitution of coefficients a_{ij} ($i = 1,2; j = 0, 1, 2$) from Eqs. (7.3.3) and (7.3.4) into the system (7.3.2) allows calculation of the bending moment M_0 and the additional thrust force H_0 at the perimeter of the dome. It should be noted that these thrust forces' values should be algebraically added to the total thrust forces in the dome (in the membrane state).

7.4 Numerical examples for domes

7.4.1 Hemispherical statically determined dome

Given a hemispherical RC dome ($\varphi_0 = 90°$) with a diameter D = 40 m, concrete class is C 20. The dome thickness $h = D/50$. The dome is subjected to the following loads: dead load is $\Delta g_k = 1$ kN/m² and live load, uniformly distributed along the horizontal projection of the dome, $q_{lsk} = 1.2$ kN/m².

It is required to calculate the membrane forces and to draw their diagrams.

Solution

The radius of the dome is:

$$R = D/2 = 40/2 = 20 \text{ m}$$

$$r_0 = R = 20 \text{ m}$$

and the dome thickness is:

$$h = D/500 = 40/500 = 0.08 \text{ m}$$

Self-weight and dead load

$$g_d = (g_k + \Delta g_k)\,1.4 = (0.08 \cdot 24 + 1)\,1.4 = 4.1 \text{ kN/m}^2$$

The calculations are performed according to Eqs. (7.2.10)–(7.2.14) and Figure 7.2.2a. At the dome vertex

$$N_1 = -0.5\,g_d\,R = -0.5 \cdot 4.1 \cdot 20 = -41 \text{ kN/m}$$

$$N_2 = N_1 = -41 \text{ kN/m}$$

$$H = -N_1 = 41 \text{ kN/m}$$

Along the perimeter of the bottom ring

$$N_1 = -g_d\,R = -4.1 \cdot 20 = -82 \text{ kN/m}$$

$$N_2 = -N_1 = 82 \text{ kN/m}; \qquad \varphi_{N2\,=\,0} = 51.83°$$

$$H = 0$$

Live load

$$q_d = 1.6\,q_k = 1.6 \cdot 2 = 3.2 \text{ kN/m}^2$$

According to Eqs. (7.2.16)–(7.2.19) and Figure 7.2.2b, at the dome vertex

$$N_1 = -0.5 \, q_d \, R = const = -0.5 \cdot 3.2 \cdot 20 = -32 \text{ kN/m}$$

$$N_2 = N_1 = -32 \text{ kN/m}$$

$$H = -N_1 = 32 \text{ kN/m}$$

$$\varphi_{N2=0} = 45°$$

Along the perimeter of the bottom ring

$$N_2 = -N_1 = 32 \text{ kN/m}$$

$$H = 0$$

$$\varphi_{N2=0} = 45°$$

Snow load

$$q_{sd} = 1.3 \, q_{sk} = 1.3 \cdot 1.2 = 1.56 \text{ kN/m}^2$$

According to Eqs. (7.2.20)–(7.2.25) and Figure 7.2.2c, at the dome vertex

$$N_1 = -0.5 \, q_{sd} \, R \cos 0° = const = -0.5 \cdot 1.56 \cdot 20 = -15.6 \text{ kN/m}$$

$$N_2 = N_1 = -15.6 \text{ kN/m}$$

$$H = -N_1 = 15.6 \text{ kN/m}$$

$$\varphi_{N2=0} = 45° \text{ and } 90°$$

$$N_{2\,max} = +0.136 \, q_{sd} \, R = 0.136 \cdot 1.56 \cdot 20 = 4.24 \text{ kN/m}$$

$$\varphi_{N2\,max} = 65.9°$$

Along the perimeter of the bottom ring

$$N_1 = N_2 = H = 0$$

7.4.2 Calculation of the local bending moment

For the same data, like in the previous example, it is required to calculate the local moment near the supporting ring.

Solution

The calculations are performed according to Eqs. (7.2.36)–(7.2.42) and Figure 7.2.3b, c. The total load that acts on the dome is:

$$q_{d\,tot} = g_d + q_d + q_{sd} = 4.1 + 3.2 + 1.56 = 8.86 \approx 8.9 \text{ kN/m}^2$$

$$\xi \approx 1.3 \, (R/h)^{1/2} = 1.3 \, (20/0.08)^{1/2} = 20.6$$

$$N_{20\,tot} = N_{20\,g} + N_{20\,q} + N_{20\,qs} = 82 + 32 + 0 = 114 \text{ kN/m}$$

$$M_{loc} = 0.295 \, N_{20\,tot} \, h = 0.295 \cdot 114 \cdot 0.08 = 2.69 \text{ kN m/m}$$

As $N_{20\,tot} > 0$, the local moment is positive, i.e., it corresponds to Figure 7.2.3c. The local moment's sign varies. The maximum value of the moment with an opposite sign, i.e., negative moment, appears at distance x from the support:

$$x = \frac{\pi \, r_0}{2 \, \zeta} = \frac{\pi \cdot 20}{2 \cdot 20.6} = 1.52 \text{ m}$$

$$|M_{min}| = 0.208 \, |M_{max}| = 0.208 \cdot 2.69 = 0.56 \text{ kN m/m}$$

7.4.3 Calculation of the supporting ring of a pre-cast dome

For the same data, like in example 7.4.1, it is required to calculate the total thrust force from the dome, the tensile force and the required reinforcement section for the supporting ring.

Solution

According to Eqs. (7.2.28)–(7.2.34) and Figure 7.2.3a,

$$H_{tot\,g} = 0.57 \, g_d \, R = 0.57 \cdot 4.1 \cdot 20 = 46.74 \text{ kN/m}$$

$$H_{tot\,q} = 0.5 \, q_d \, R = 0.5 \cdot 3.2 \cdot 20 = 32 \text{ kN/m}$$

$$H_{tot\,qs} = 0.39 \, q_{sd} \, R = 0.39 \cdot 1.56 \cdot 20 = 12.17 \text{ kN/m}$$

$$\Sigma H_{tot} = H_{tot\,g} + H_{tot\,q} + H_{tot\,qs} = 46.74 + 32 + 12.17 = 90.91 \text{ kN/m}$$

$$T = \Sigma H_{tot} \, r_0 = 90.91 \cdot 20 = 1818.2 \text{ kN}$$

If there is no pre-stress in the supporting ring, then the required reinforcement section is:

$$A_{sT} = T/f_{sd} = 1818.2 \cdot 10^3 / 350 = 5195 \text{ mm}^2 = 51.95 \text{ cm}^2$$

Generally, in large domes, as the discussed one, pre-stressed reinforcement is used. Assuming that the design strength of the pre-stressed steel is $f_{spd} = 900$ MPa,

$$A_{spT} = T/f_{spd} = 1818.2 \cdot 10^3 / 900 = 2020 \text{ mm}^2 = 20.2 \text{ cm}^2$$

7.4.4 Calculation of the elastic connection between the dome and the supporting ring

Given a spherical dome that corresponds to the following data:

$R = 23$ m, $r_0 = 15$ m, $\varphi_0 = 40°$, $h = 0.06$ m, $b_r = 0.6$ m, $h_r = 0.5$ m, $e = h_r/2 = 0.25$ m

The total design load that acts on the dome is:

$$q_{d\,tot} = 6 \text{ kN/m}^2$$

It is required to calculate the elastic connection between the dome and the supporting ring to find the additional thrust force and moment.

Solution

Following Eqs. (7.3.3), (7.3.4) and Figure 7.3.1 and substituting cylindrical stiffness instead of D in Eq. (7.3.5), yields:

$$\sin \varphi_0 = 0.644, \quad \cos \varphi_0 = 0.765$$

$$s = 0.76 \, (R \, h)^{1/2} = 0.76 \, (23 \cdot 0.06)^{1/2} = 0.893 \text{ m}$$

$$D = E_c \, h^3/12 = E_c \, 0.06^3/12 = 1.8 \cdot 10^{-5} \, E_c$$

$$r_0 = R \sin \varphi_0 = 23 \cdot 0.644 = 14.812 \text{ m}$$

$$E_c \, a_{11} = \frac{12 \, s}{h^3} + \frac{12 \, r_0^2}{b_r \, h_r^3} = \frac{12 \cdot 89.3}{6^3} + \frac{12 \cdot 1481.2^2}{60 \cdot 50^3} = 8.458 \; 1/\text{cm}^2$$

$$E_c \, a_{12} = E_c \, a_{21} = \frac{6 \, s^2}{h^3} \sin \varphi_0 - \frac{12 \, r_0^2}{b_r \, h_r^3} \, e = \frac{6 \cdot 89.3^2}{6^3} \, 0.644 - \frac{12 \cdot 1481.2^2}{60 \cdot 50^3} \, 25 = 55 \; 1/\text{cm}$$

$$E_c \, a_{22} = \frac{6 \, s^3}{h^3} \sin^2 \varphi_0 + \frac{r_0^2}{b_r \, h_r} + \frac{12 \, r_0^2}{b_r \, h_r^3} \, e^2 =$$

$$= \frac{6 \cdot 89.3^3}{6^3} \, 0.644^2 + \frac{1481.2^2}{60 \cdot 50} + \frac{12 \cdot 1481.2^2}{60 \cdot 50^3} \, 25^2 = 11087.4$$

$$a_{10} = \frac{q \, s^4}{2 \, R \, D} \sin \varphi_0 ; D = 1.8 \; 10^{-5} \, E_c$$

$$E_c \, a_{10} = \frac{q \, s^4}{2 \, R \, D} \sin \varphi_0 = \frac{6 \cdot 0.893^4}{2 \cdot 23 \cdot 1.8 \, 10^{-5}} \, 0.644 = 2962.06 \text{ kN/m}^2 = 296.206 \text{ N/cm}^2$$

$$E_c \, a_{20} = \frac{q_{d\,tot} \, R \, r_0}{h} \, (\frac{1}{1 + \cos \varphi_0} - \cos \varphi_0) \sin \varphi_0 + \frac{q_{d\,tot} \, R \, r_0^2}{b_r \, h_r} \frac{\cos \varphi_0 \, \sin^2 \varphi_0}{1 + \cos \varphi_0} =$$

$$= q_{d\,tot} \, R \left[\frac{r_0}{h} \left(\frac{1}{1 + \cos \varphi_0} - \cos \varphi_0 \right) \sin \varphi_0 + \frac{r_0^2}{b_r \, h_r} \frac{\cos \varphi_0 \, \sin^2 \varphi_0}{1 + \cos \varphi_0} \right] =$$

$$= 6\,23 \left[\frac{14.812}{0.06}\,(\frac{1}{1+0.765}-0.765)\,0.644 + \frac{14.812^2}{0.6\,0.5}\,\frac{0.765\,0.644^2}{1+0.765} \right] =$$

$$138\,[-\,31.599 + 131.64] = 13805.658\ \text{kN/m} = 138{,}056.58\ \text{N/cm}$$

Substitution of these values in the system (7.3.2) yields:

$$8.458\ M_0 + 55\ H_0 + 296.206 = 0$$

$$55\ M_0 + 11087.4\ H_0 + 1380.566 = 0$$

Consequently,

$$H_0 = -\,12.54\ \text{N/cm} = -\,1.254\ \text{kN/m}$$

$$M_0 = -\,46\ \text{N cm/cm} = -\,0.046\ \text{kN m/m}$$

7.4.5 Calculation of the dome critical buckling load

Let's calculate the buckling load for the Aqua Park dome in Moscow (see corresponding data in Section 7.1.2). Following Eq. (4.2.24),

$$q_{cr} = 0.2\ E_c\,(t)\,(h_m/R_1)^2$$

where

$$E_{c.red} = 0.319 \cdot 34500 = 11006\ \text{MPa}; \quad h_m = 25\ \text{cm};\ R_1 = 163.8\ \text{m}$$

Consequently,

$$q_{cr} = 0.2 \cdot 11006\,(0.25/163.8)^2 = 0.00513\ \text{MPa} = 5.13\ \text{kN/m}^2$$

The design load for the real shell, including the snow load of 1 kN/m² (the structure collapsed in February), is 7.0 kN/m². It is evident that even in the shell zone with rather big curvature the structure is able to take just 73% of its design load. In the bottom zone that has low curvature the situation is even more critical. The main problem is geometrically non-linear behavior of a thin-walled shell under long-term loads (concrete creep) that should be considered in the design process (Ishakov and Ribakov 2014).

8

Investigations of a Full-Scale RC Dome Under Vertical Vibrations

8.1 General

Monitoring and analysis of natural vibration frequencies of long-span structures can provide fast and reliable results about their real state. Testing a structure under vibration loading can provide valuable information about its dynamic parameters, like natural frequencies, vibration modes and damping ratios. This information can be further used to calibrate theoretical models, develop modeling techniques, and verify theoretically predicted damage.

Sometimes the structural assessment can include non-destructive testing and additionally finite element modeling. Based on the obtained results, proper methods for retrofitting of existing long span structures damaged by earthquakes, impact and other dynamic excitations can be developed.

As known, if a long span structure is constructed on rather soft soils, the seismic wave length can be of the same order like the length of the structure. In such a case out of phase supports' vibrations can appear not only in the horizontal, but in the vertical direction too. This phenomenon has taken place, for example, in 1952 in one of the buildings analyzed and described by Housner (1957). The out of phase supports' vibrations can be superposed by different forms of natural vibrations of the structure, which can increase or decrease the overall vibrations' amplitudes. Therefore the phenomenon should be taken into account in design of long span structures.

The authors have investigated experimentally and theoretically, the dynamic response of a spherical RC dome in Dushanbe (Tajikistan), shown in Figure 8.1.1. The structure was subjected to vertical dynamic loads, generated by a vibration machine. The experimental results were compared with analytical ones and with those of finite elements analysis.

Figure 8.1.1. A general view of the tested dome.

8.2 Analytical investigation of long span shells due to out of phase supports' vibrations

As many shells are long span structures, it is proposed to consider in analysis the ratio between the length of the span and that of the seismic wave. This can be done by introducing a coefficient $\eta_i\,(x, y)$ that takes into account possible out of phase supports' vibrations (x and y are the dimensions of the shell in plane as shown in Figure 8.2.1).

This coefficient depends on the length of the seismic wave in soil, λ, as follows (Housner 1957):

$$\lambda = c\,T_0 \tag{8.2.1}$$

where c is the seismic wave velocity, m/sec;

T_0 is the structure's dominant vibration period, sec.

If $\lambda \geq L$,

$$\eta_i\,(x,\,y) = \frac{\int_0^{Lx} \int_0^{Ly} m\,(x,y)\Phi_i\,(x,y)dx\,dy}{\int_0^{Lx} \int_0^{Ly} m\,(x,y)\Phi_i^2\,(x,y)dx\,dy} \tag{8.2.2}$$

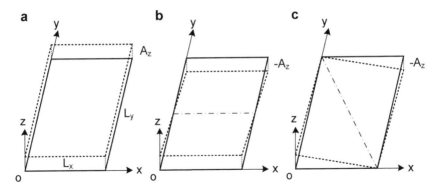

Figure 8.2.1. Vertical vibrations of long span shells (the shell itself is not shown): a—vertical in phase displacements; b—opposite sides out of phase displacements; c—similar to case "b" in diagonal direction.

If $\lambda < L$,

$$\eta_i (x, y) = \frac{\int_0^{Lx} \int_0^{Ly} m\ (x,y)\mu(x,y)\ \Phi_i\ (x,y)dx\ dy}{\mu(x,y) \int_0^{Lx} \int_0^{Ly} m\ (x,y)\Phi_i^2\ (x,y)dx\ dy} \tag{8.2.3}$$

In these formulas:

$\Phi_i\ (x, y)$ is the i-th form of the structure's natural vibration mode;

$m\ (x, y)$ is the mass of a unit area;

L_x, L_y are the structure's dimensions in plane;

$\mu\ (x, y)$ is the distribution function of supports' vibrations in plane (in or out of phase).

In a case when $\lambda \le L$ has an especial interest the function $\mu\ (x, y)$ creates a new surface that is deformed in the vertical direction under the dynamic load due to supports' vibrations. This surface, in turn, vibrates according to mode $\Phi_i\ (x, y)$, as shown in Figure 8.2.1.

Let's define the $\mu\ (x, y)$ function for each of the cases shown in this figure. An equation of a conventional plane that is jointed with the XOY is $z = 0$. Joining the μ function with the Z axis, yields the following equation of the shell surface vibration:

$$\frac{x}{a} + \frac{y}{b} + \frac{\mu}{A_z} = 1 \tag{8.2.4}$$

where a, b and A_z are lengths of axes X, Y and Z, cut by the plane (in Figure 8.2.1 a and b are not shown as their values are significantly lower, compared to A_z).

Assuming that the shell has a rectangular plane (see Figure 8.2.1a),

$$\mu_1 (x, y) = A_z = \text{const} \tag{8.2.5}$$

where A_z is the peak amplitude of the shell support.

Similarly, according to Figure 8.2.1b and c,

$$\mu_2 (x,y) = A_z (1 - 2\, y/L_y) \tag{8.2.6}$$

$$\mu_3 (x, y) = A_z (1 - x/L_x - y/L_y) \tag{8.2.7}$$

In long span shells $L_x \ll L_y$, therefore the second term in the brackets in Eq. 8.2.7 can be ignored and

$$\mu_2 = \mu_3 \tag{8.2.8}$$

Considering that the natural vibration mode of the shell in the long direction, y, has a sinusoidal form,

$$\Phi_i (y) = \sin (i\, \pi\, y/L_y),\ i = 1, 2, 3, \ldots, \text{ and } m\, (x,y) = \text{const} \tag{8.2.9}$$

In a case when $\lambda \geq L_y$, the coefficient $\eta_i (x, y)$ takes the form:

$$\eta_i (y) = \frac{\int_0^{L_y} \Phi_i (y)\, dy}{\int_0^{L_y} \Phi_i^2 (y)\, dy} \tag{8.2.10}$$

Analyzing separately the numerator and the denominator of Eq. (8.2.10) it is possible to show that:

$$\int_0^{L_y} \Phi_i (y)\, dy = -\frac{L_y}{i\,\pi} (\cos i\,\pi - 1) \tag{8.2.11}$$

and

$$\int_0^{L_y} \Phi_i^2 (y)\, dy = \frac{L_y}{i\,\pi} (- 0.25 \sin 2i\,\pi + 0.5\, i\,\pi) = 0.5\, L_y \text{ for any } i \tag{8.2.12}$$

For $i = 1$: $\eta_1 (y) = 4/\pi$; $i = 2$: $\eta_2 (y) = 0$; $i = 3$: $\eta_3 (y) = 4/(3\pi$ and so on. For any even value of i $\eta_i = 0$.

In a case when $\lambda < L$:

$$\eta_i (y) = \frac{\int_0^{L_y} \mu_i\, \Phi_i (y)\, dy}{\mu_i \int_0^{L_y} \Phi_i^2 (y)\, dy} \tag{8.2.13}$$

Assuming that $A_z = 1$, $i = 1$: $\mu_1 = 1 = \text{const}$, $\eta_1 (y) = 4/\pi$. For $i = 2$ and $i = 3$: $\mu_2 (y) = \mu_3 (y) = 1 - 2\, y/L_y$.

Let's define:

$$y^* = i\, \pi\, y/L_y \tag{8.2.14}$$

If $y = 0$ than $y^* = 0$; if $y = L_y$ then $y^* = i\pi$.

Let's analyze separately the numerator of Eq. (13):

$$\int_0^{L_y} (1 - \frac{2y}{L_y}) \text{Sin} \frac{i\pi y}{L_y} dy = \int_0^{L_y} \text{Sin} \frac{i\pi y}{L_y} dy - \frac{2}{L_y} \int_0^{L_y} y \text{Sin} \frac{i\pi y}{L_y} dy =$$

$$= \frac{2 L_y}{\pi} - \frac{2 L_y}{i^2 \pi^2} \int_0^{i\pi} y^* \text{Sin } y^* dy^* \qquad (8.2.15)$$

where

$$\int_0^{i\pi} y^* \sin y^* dy^* = [-y^* \cos y^* + \int_0^{i\pi} \cos y^* dy^*]_0^{i\pi} =$$

$$= \{-y^* \cos y^* + [\sin y^*]_0^{i\pi}\}_0^{i\pi} = [-y^* \cos y^*]_0^{i\pi} + \sin i\pi = i\pi \cos i\pi + \sin i\pi$$

$$(8.2.16)$$

Then

$$\frac{2 L_y}{\pi} - \frac{2 L_y}{i^2 \pi^2}]_0^{i\pi} \sin y^* dy^* = \frac{2 L_y}{\pi} [1 - \frac{1}{i^2 \pi} (-i \pi \cos i\pi + \sin i\pi)] \quad (8.2.17)$$

Let's analyze the denominator of Eq. (8.2.13), taking into account that

$$\mu_2 (y) = (1 - 2 y/L_y) \, 0.5 \, L_y; \, 0 \le y \le 0.5 \, L_y \qquad (8.2.18)$$

Substituting the above obtained expressions into Eq. (8.2.13) yields:

$$\eta_i (y) = 4 \{1 - [1/(i^2 \pi)] [- i \pi \cos i\pi + \sin i\pi]]\} / [\pi (1 - 2 y/L_y)] \qquad (8.2.19)$$

Substituting the coordinates of the supports $y = 0$; $y = L_y$ into Eq. (8.2.19), yields:

for $i = 1$, $\eta_1 (y) = 0$; $i = 2$, $\eta_2 (y) = 6/[\pi (1 - 2 y/L_y)]$.

If $y = 0$ then $\eta_2 = 6/\pi$; at $y = L_y$ $\eta_2 = -6/\pi$.

For $i = 3$, $\eta_3 (y) = 8/[3 \pi (1 - 2 y/L_y)]$.

If $y = 0$ then $\eta_3 = 8/(3 \pi)$; at $y = L_y$ $\eta_3 = -8/(3 \pi)$.

Thus, based on analytical results, numerical values of coefficient η for different vibration modes, i, of the shell can be received. This coefficient is multiplied by a vertical dynamic load acting on the shell, and in such a way, the design load is obtained more accurately. Among all obtained values of η the maximum absolute value should be selected as design coefficient for out of phase supports' vibrations.

For example, Let's discuss a long span shell with a span of 100 m, assuming rather soft soil under foundations, i.e., $\lambda_w < L$, where λ_w is the length of the dynamic load waves. In this case for $i = 1$, and $y = 0$ or $y = L_y$ the corresponding $|\eta_{max}| = 6/\pi$. The coefficient η that corresponds to $i = 3$ and the same y values will be lower: $|\eta_3| = 8/(3 \pi)$.

8.3 Vibration testing of the RC dome

A dome of a public building in Dushanbe (Tajikistan) was tested under dynamic loading, applied by a vibration machine. The general view of the tested structure is presented in Figure 8.1.1 above.

The dome is a shallow thin-walled spherical RC shell. Its outside diameter is 30 m and the inside one is 26.5 m, the supporting ring width is 1.75 m. A structural scheme of the RC dome is shown in Figure 8.3.1. The dome rise is 3.6 m and its thickness is 6 cm. It includes a system of meridian and parallel ribs. The meridian ribs limit the sector with a central angle of 12° and come to the upper ring with a diameter of 3.3 m. This ring has a 19 cm section height (excluding the dome thickness) and a 20 cm width rib along the perimeter. This ring is reinforced by two longitudinal periodic shape steel rods with a diameter of 14 mm and links with a diameter of 8 mm each 18 cm. The part of the dome located inside this ring is reinforced by a net, made of 8 mm diameter steel rods which are each placed 20 cm in both directions.

The meridian ribs of the dome have a variable height (from 14 to 44 cm), increasing in the supporting ring direction. The ribs' width is 8 cm. These ribs are reinforced by two 25 mm periodic shape rods and 8 mm links, placed with a 10–15 mm step. The parallel rings have a step of 1.5 m and a section of 8 x 8 cm. These rings are reinforced by 18 mm periodic shape rods and 8 mm links, placed with a step of 20 cm.

The dome itself is reinforced between the ribs by a net, made of 6 mm diameter steel rods placed with a 15 cm step. The supporting ring is reinforced by a net of 8 mm steel rods and a spatial frame made of steel rods with a diameter of 28 mm and a 10 cm step along the height of the ring. The height of the supporting ring varies from 10 to 60 cm.

The dome is supported on 15 columns, located along the perimeter, as shown in Figure 8.3.1. The structure is designed for a zone with seismic hazard corresponding to degree 8 according to the Modified Mercally scale. Therefore, experimental investigation of the dome's dynamic parameters (natural vibration modes, corresponding periods and damping ratios) in vertical and horizontal directions is important for assessment of the structure seismic resistance.

The dynamic load was applied to the shell by a 500 kg vibration machine (VM), located at the dome vertex. The VM caused vertical and horizontal vibrations of the structure. The frequency of the applied dynamic load was changed smoothly to cause resonant vibrations, allowing identification of the dome's natural dynamic parameters.

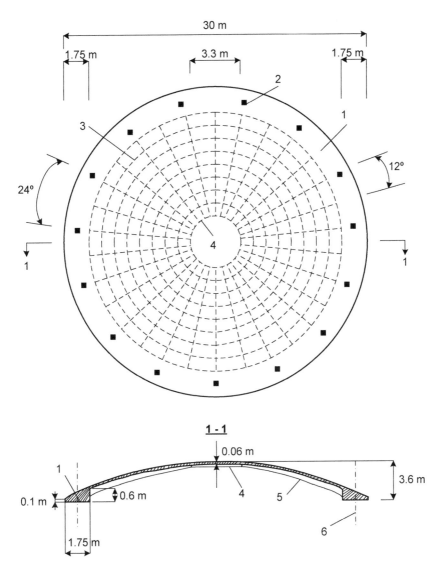

Figure 8.3.1. A structural scheme of the investigated dome: 1—supporting ring; 2—columns; 3—parallel ribs; 4—upper ring; 5—meridian ribs; 6—column axes.

For recording the vibrations of the dome, velocities and displacements were measured. A scheme of the measuring equipment location is shown in Figure 8.3.2a. Two types of devices were used in the experiments: K001 for measuring accelerations and OSP for velocities. The dynamic parameters of the dome were measured in X and Z directions.

The eccentric masses, used in the VM, were selected so that nonlinear behaviour of the dome would be prevented. Only the first vibration mode was identified in the horizontal direction (Figure 8.3.2b) and two symmetrical modes in the vertical direction (Figure 8.3.2c).

Table 8.3.1 presents the results of dynamic testing.

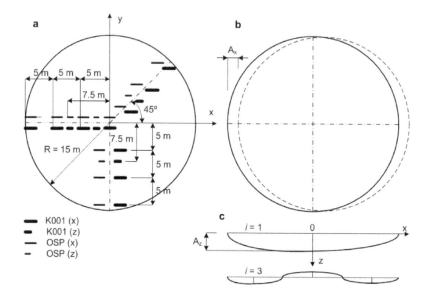

Figure 8.3.2. Location of measuring equipment on the dome (a); natural mode shapes in horizontal (b) and vertical (c) directions.

Table 8.3.1. Experimentally obtained natural dynamic parameters of the dome for the first natural vibration modes in horizontal and vertical directions.

Dynamic parameter	Horizontal direction	Vertical direction
Natural period, sec	0.22	0.106...0.22
Damping ratio	0.07	0.04...0.10
Peak displacement, mm	0.25	0.10...0.37
Peak velocity, cm/sec	0.94...0.98	0.63...2.08

Following the obtained results, the vibrations of the structure in the horizontal direction can be simulated as for a single degree of freedom system. In other words, the RC spherical shell itself is stiff enough in the horizontal direction, in spite that it is shallow and thin-walled. At the same time, in the vertical direction the first two symmetrical natural vibration modes were identified. It shows that this direction is the design one. Proving this fact was the main goal of this study that was focused on vertical vibrations of long span structures.

8.4 Analytical investigation of RC domes vertical vibrations

Reisner (1946) has performed one of the first analytical investigations on shallow spherical shells' natural frequencies and mode shapes. Practical aspects in dynamic theory of RC shells were further developed (Oniashivili 1957) and applied for seismic design (Bobrov et al. 1974).

Following these approaches, the expression for calculating the natural vibration frequencies was proposed:

$$\omega_{mn}^2 = [g/(\gamma h)] [D (\lambda_n^2 + \mu_m^2)^2 + E h/R^2] \tag{8.4.1}$$

where m and n are number of half-waves in the vibration modes in X and Y directions, respectively;

R is the shell radius of curvature;

D is the cylindrical stiffness of a shell that is obtained as follows:

$$D = E h^3/[12 (1 - v^2)] \tag{8.4.2}$$

Here E is the elasticity modulus of concrete;

h is the shell thickness;

v is the Poisson's coefficient.

Additionally,

$$\lambda_n = n \pi/d \tag{8.4.3}$$

$$\mu_m = m \pi/d \tag{8.4.4}$$

where d is the dome diameter.

It can be shown that the first term, $D (\lambda_n^2 + \mu_m^2)^2$, in Eq. (8.4.1) presents the contribution of bending, whereas the second term, $E h/R^2$, presents the contribution of membrane forces. It will be further shown that for thin walled shells the first term is negligible, compared to the second one.

Let's analyse the above-discussed spherical RC dome using Eq. (8.4.1). Taking into account that the shell is spherical, it will vibrate according to symmetric modes only and therefore $n = m = 1$, 3, etc. Considering that for concrete Poisson coefficient $v_c^2 \ll 1$, Eq. (8.4.1) takes the following form:

$$\omega_{mn}^2 = [g/(\gamma h)][(E_c h^3/12)(2 \lambda_n^2)^2 + E_c h/R^2] = [(g E_c/\gamma)][(h^2/3) \lambda_n^4 + 1/R^2] \tag{8.4.5}$$

For $n = m = 1$ and $1/R^2 = 64 f^2/d^4$, where f is the shell's height.

$$\omega_{11}^2 = (g\, E_c/(\gamma\, d^4))\, [(\pi^4\, h^2/3) + 64\, f^2] = [g\, E_c/(\gamma\, d^4)]\, (32.47\, h^2 + 64\, f^2) \tag{8.4.6}$$

For thin-walled shells $h^2 \ll f^2$. Therefore $\omega_{11}^2 = 64 f^2 E_c g/(\gamma d^4)$ and in this case:

$$\omega_{11} = (8 f/d^2)\, (E_c\, g/\gamma)^{0.5} = k/(E_c\, g/\gamma)^{0.5} \tag{8.4.7}$$

where $k = 1/R$ is the shell curvature.

Correspondingly, the natural vibration period:

$$T_{11} = 2\,\pi/\omega_{11} = [\pi\, d^2/(4 f)]\, [\gamma/(E_c\, g)]^{0.5} \tag{8.4.8}$$

Let's define

$$c = 0.25\,\pi\, [\gamma/(E_c\, g)]^{0.5},\ \sec/m \tag{8.4.9}$$

Then $1/c$, m/sec is a parameter of dynamic waves' velocity in concrete. Therefore,

$$T_{11} = c\, d^2/f \tag{8.4.10}$$

As known, the concrete modulus of elasticity depends on concrete creep and the surrounding environment's temperature:

$$T_{11} = f\, (E_{c\,red}) \tag{8.4.11}$$

where $E_{c\,red}$ is the reduced value of concrete elasticity modulus, taking into account the above-mentioned factors (Iskhakov and Ribakov 2013).

Therefore,

$$T_{11} = 2\,\pi/\omega_{11} = [\pi\, d^2/(4 f)]\, [\gamma/(E_{c\,red}\, g)]^{0.5} \tag{8.4.12}$$

For $n = m = 3$

$$\omega_{33}^2 = [g\, E_{c\,red}/(\gamma\, d^4)]\, (2630\, h^2 + 64\, f^2) \tag{8.4.13}$$

Let's compare the analytical and experimental results for the investigated dome ($d = 30$ m, $f = 3$ m, $\gamma = 24$ kN/m^3, $g = 9.81$ m/s^2, $E_c = 23800$ MPa, $h_m \approx 10$ cm). Following Iskhakov and Ribakov (2013), $E_{c\,red} = 0.375\, E_c$. For $n = m = 1$, the corresponding natural vibration period following Eq. (8.4.12) is:

$$T_{11} = [3.14\, 30^2/(4\, 3)]\, [24/(0.375\, 23800\, 10^3\, 9.81)]^{0.5} = 235.5\, 5.185\, 10^{-4} = 0.122\ \text{s}.$$

For $n = m = 3$, following Eq. (8.4.13),

$$2630\, h^2 + 64\, f^2 = 10^5\, (2.63 + 57.6)$$

Similar to the case, when $n = m = 1$, the first term is negligible, compared to the second (2.63 « 57.6), therefore, the first term can be neglected. Hence,

$$\omega_{33} \approx \omega_{11}, \text{ or } T_{33} \approx T_{11} = 0.122 \text{ sec}$$

The measured natural vibration period value $T_{11 \, exp} = 0.106...0.220$ sec (see Table 8.3.1). The theoretical value is within the experimentally obtained range.

8.5 Finite elements analysis of the dome natural vibration period

The finite elements (FE) analysis was aimed at numerical modelling of the investigated spherical RC dome, excited by a VM. This analysis was performed by the authors together with Professor G.G. Kashevarova (Perm National Research Polytechnic University, Russia) and her research group. The numerical results are compared with the experimentally and analytically dynamic parameters of the dome that were obtained before.

ANSYS software was used for the dome analysis. SOLID65 element with three degrees of freedom was used for creating the FE model. The model of the dome includes 286785 nodes and 209836 elements. The supporting columns have a section of 40 x 40 cm and are located along the perimeter of the dome each 24°.

Following the results of FE analysis, the first mode natural vibration period in the vertical direction is 0.131 sec. The experimental value is 0.106...0.220 sec (see Table 8.3.1) and the analytical result is 0.122 sec. The results obtained by all methods (experimental, analytical and FE analysis) are very close. As the analytical approach is rather simple, it can be used for designing such RC domes.

9

Long Reinforced Concrete Folders

9.1 Constructive requirements

9.1.1 General

Folders are structures composed of rather thin plates and aimed to cover a big span area without columns (Figure 9.1.1a). A folder is considered to be long if its length, l_1, is more than the width, l_2, and

$$l_2 \leq 0.5\, l_1 \qquad\qquad (9.1.1)$$

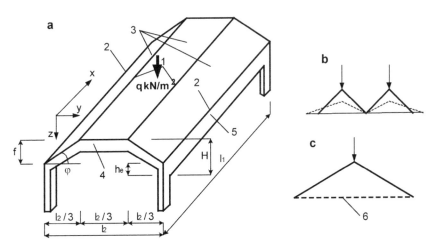

Figure 9.1.1. A structure of a folder: a—general view; b—thrust shape; c—tensile bar (6); 1, 2, 3, 4 and 5—upper ribs, lower ribs, diaphragm (load carrying wall, bar), edge beams, respectively.

A folder enables to reduce the quantity of materials (relative to ordinary roofs) and consists of plates and beams. A folder is an effective structure, if it includes such constructive elements as diaphragms, bars or load carrying walls at both edges. These elements prevent thrusting of the folder (Figure 9.1.1b, c) and take the corresponding thrust forces that appear at the edges. The load carrying walls are vertical supports for the folder, which in this case replace diaphragms.

Edge elements should be constructed in the longitudinal direction of the folder. It is necessary to provide the out-of-plane stiffness of the folder's thin plates (position 5 in Figure 9.1.1a).

9.1.2 Main dimensions

The spans l_1 and l_2 of the folder are usually known, and given by the architect. Additionally, the folder width is practically limited to:

$$l_2 \leq 6 \text{ m} \tag{9.1.2}$$

Other dimensions are defined by constructive requirements that are based on engineering experience. The folder thickness:

$$h = (1/100 \dots 1/200)\, l_2; \quad 5 \text{ cm} \leq h \leq 10 \text{ cm} \tag{9.1.3}$$

The total folder rise is:

$$H = (1/10 \dots 1/15)\, l_1 \tag{9.1.4}$$

In shallow folders

$$f = (1/6 \dots 1/8)\, l_2 \tag{9.1.5}$$

The edge element's section has the following dimensions:

$$h_e = (0.2 \dots 0.4)\, H; \quad b_e = (0.2 \dots 0.4)\, h_e \tag{9.1.6}$$

9.2 The folder action in the long direction

9.2.1 Approximate calculation of a folder using an equivalent section

A folder behaves as a simple supported beam in its long direction, l_1, i.e., it is in ordinary bending. The main design assumptions are:

- the transverse section of the folder is non-deformed;
- the section reaches its plastic stage like in ordinary RC beams.

Based on the first assumption, the folder section that has a form of a triangle or trapezoidal can be replaced by an equivalent inverse T or I section (Figure 9.2.1a, b).

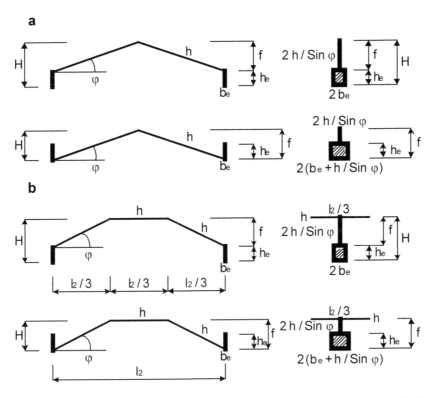

Figure 9.2.1. Given folder sections and their equivalent ones in a form of inverse T (a) and I (b).

Calculation of folder in bending in the long direction is performed as for a beam with an equivalent section. The section dimensions are selected according to Figure 9.2.1. The thickness of the equivalent section's wall is:

$$h_{eq} = \frac{2h}{Sin\ \varphi}$$ (9.2.1)

The width of the bottom flange in the equivalent section is:

$$b_{e\ eq} = 2\ b_e$$ (9.2.2)

If it is necessary to provide water drainage from the roof, a channel is constructed between the folder and the edge element (i.e., the edge element will be above the bottom plate of the folder):

$$b_{e\,eq} = 2\,(b_e + \frac{h}{Sin\,\varphi}) \qquad (9.2.3)$$

The bending moments and shear forces under the external loads are calculated, like for simple supported beams:

$$M_d = \frac{F_d\,l_2\,l_1^2}{8}\,kN\,m;\quad V_d = \frac{F_d\,l_2\,l_1}{2}\,kN \qquad (9.2.4)$$

9.2.2 Exact calculation of a folder with a triangular section

For exact calculation of a folder with a triangular section, like for long cylindrical shells (Chapter 5), it is assumed that, when deflections occur, the folder has a non-deformed section. As for ordinary RC structures, the calculations can be aimed at:

- checking the folder with a given reinforcement section, A_s, in the edge elements (case 1), or
- calculating the reinforcement section, A_s, for a given design moment M_d that acts in the folder (case 2).

In both cases the same calculation scheme of the folder section is used (Figure 9.2.2).

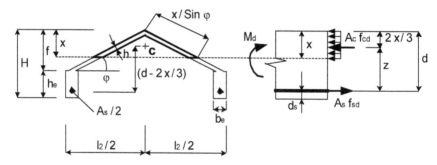

Figure 9.2.2. Calculation scheme of a folder with triangular section.

Checking the section (case 1)

The compressed section zone area is:

$$A_c = 2\,h\,x/\sin\,\varphi \qquad (9.2.5)$$

From the equation of forces' equilibrium follows:

$$A_c f_{cd} = A_s f_{sd} \qquad (9.2.6)$$

Hence, the compressed zone depth is:

$$x = \frac{A_s\, f_{sd}\, \sin \varphi}{2\, h\, f_{cd}} \tag{9.2.7}$$

where $\varphi = \arctan f/(l_2/2)$

The internal forces lever is:

$$z = d - 2\, x/3 \tag{9.2.8}$$

The compressed section zone area should satisfy the following requirement (En 1992):

$$S_c/S_0 \le 0.64 \tag{9.2.9}$$

where

S_c is the static moment of the compressed section zone (its center of gravity is at point c in Figure 9.2.2) relative to an axis that passes through the center of gravity of the tensile reinforcement, A_s:

$$S_c = \frac{2\, h\, x}{\sin \varphi}\left(d - \frac{2}{3}x\right) \tag{9.2.10}$$

S_0 is the static moment of the equivalent section relative to the same axis:

$$S_0 = \frac{2\, h\, f}{\sin \varphi}\left(d - \frac{2}{3}f\right) + b_e\left(h_e - d_s\right)^2 \tag{9.2.11}$$

The moment carrying capacity of the section is:

$$M_{pl} = A_s f_{sd}\, z \tag{9.2.12}$$

The checked condition for bending is:

$$M_{pl} \ge M_d \tag{9.2.13}$$

Calculation of the section (case 2)

The moments' equilibrium equation relative to axis A_s is:

$$M_d = \frac{2\, h\, x}{\sin \varphi}\, f_{cd}\left(d - \frac{2}{3}x\right) = \frac{2\, h\, d^2}{\sin \varphi}\, f_{cd}\, \omega\left(1 - \frac{2}{3}\omega\right) \tag{9.2.14}$$

where the relative compressed zone depth is:

$$\omega = x/d = 1.5\left(1 - \sqrt{1 - \frac{M_d\, \sin \varphi}{3\, h\, d^2\, f_{cd}}}\right) \tag{9.2.15}$$

Following (EN 1992) for any section,

$$z_{max} \leq 0.95 \, d \,\, (\omega_{min} \geq 0.1) \tag{9.2.16}$$

and ω should satisfy this condition.

The reinforcement section, A_s, calculated according to Eq. (9.2.6), is:

$$A_s = \frac{2 \, h \, x \, f_{cd}}{f_{sd} \sin \varphi} \geq A_{s \, min} \tag{9.2.17}$$

9.2.3 Example for calculating a folder as an ordinary simple supported beam

Given a folder with an isosceles triangle section. The structure is made of concrete (class C 20) reinforced by ribbed steel bars. The dimensions of the folder are:

$$l_1 = 20 \text{ m}, \quad l_2 = 5 \text{ m}, \quad h = 0.07 \text{ m}, \quad f = 1.5 \text{ m}, \quad h_e = 0.5 \text{ m},$$
$$b_e = 0.15 \text{ m}, \quad d_s = 3 \text{ cm}.$$

The dead and live loads are:

$$\Delta g_k = 0.3 \text{ kN/m}^2, \quad q_k = 0.5 \text{ kN/m}^2$$

It is required to calculate the reinforcement section in the edge element, A_s.

Solution

$$\tan \varphi = 2 f/l_2 = 2 \cdot 1.5/5 = 0.6; \quad \varphi = 30.96°; \quad \sin \varphi = 0.51$$
$$d = f + h_e - d_s = 1.5 + 0.5 - 0.03 = 1.97 \text{ m}$$

The folder section area is:

$$A_g = 2 \, h \, f/\sin \varphi + 2 \, b_e \, h_e = 2 \cdot 0.07 \cdot 1.5/\sin 30.96 + 2 \cdot 0.15 \, 0.5 = 0.558 \text{ m}^2$$

$$g_k = 0.558 \cdot 24 = 13.4 \text{ kN/m}$$

$$F_d = 1.4 \, (g_k + \Delta g_k \, l_2) + 1.6 \, q_k \, l_2 = 1.4 \, (13.4 + 0.3 \cdot 5) + 1.6 \cdot 0.5 \cdot 5 = 24.9 \text{ kN/m}$$

$$M_d = \frac{F_d \, l_1^2}{8} = \frac{24.9 \, 20^2}{8} = 1245 \text{ kN m}$$

$$\omega = 1.5 \left(1 - \sqrt{1 - \frac{M_d \sin \varphi}{3 \, h \, d^2 \, f_{cd}}} \right) = 1.5 \left(1 - \sqrt{1 - \frac{1245 \cdot 10^6 \cdot 0.51}{3 \cdot 70 \cdot 1970^2 \cdot 8.7}} \right) = 0.065 < 0.1$$

$$\omega = 0.1$$

$$A_s = \frac{2\,h\,x\,f_{cd}}{f_{sd}\sin\varphi} = \frac{2\,70 \cdot 0.11970 \cdot 8.7}{350 \cdot 0.51} = 1344.2 \text{ mm}^2 = 13.44 \text{ cm}^2$$

$$A_{s\,min} = 0.0015\,A_0 = 0.0015\,(A_g - 2\,b_e\,d_s) =$$

$$= 0.0015\,(0.558 - 2 \cdot 0.15 \cdot 0.03)\,10^4 \text{ cm}^2 = 8.24 \text{ cm}^2$$

$$A_s > A_{s\,min} \rightarrow OK$$

9.3 Considering the folder plates flexibility

9.3.1 Transverse bending moments in plates

A folder plate is an element, in which the length and width are significantly big relative to its thickness, and the internal forces act in its plane (membrane state). Under transverse bending moments, when the forces act perpendicular to its plane, this plate behaves as a table (bending state). In other words, each table can behave as a plate and vice-versa.

The folder plate behaves out-of-plane as continuous table. It is assumed that supports of this table are upper and lower ribs of the folder. A reason of such assumption is that the in-plane stiffness of the plates is sufficiently

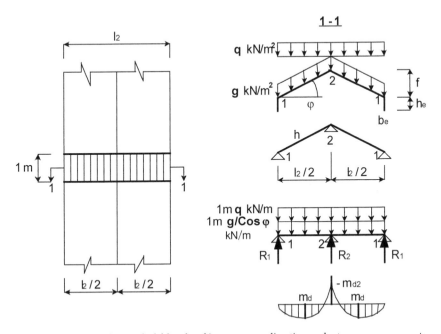

Figure 9.3.1. A static scheme of a folder chord in transverse direction under transverse moments.

higher than that in the out-of-plane direction. Therefore, the ribs' deflections are negligible relative to those of tables.

A static scheme of a folder chord with a 1 m width for calculating transverse moments is presented in Figure 9.3.1. The calculation method is similar to that for one-direction plates.

The angle φ in shallow folder is small, therefore it is possible to consider that the self-weight and the dead loads are uniformly distributed over the horizontal projection of the folder.

9.3.2 Example: calculating a folder in transverse direction

For the folder, given in Section 9.2.3, it is required to calculate the transverse moments and the support reactions.

Solution

$$\cos \varphi = \cos 30.96° = 0.858$$

The uniformly distributed load, acting on the folder chord with 1 m width is:

$$F_d = 1 \text{ m } \{[1.4 \ (0.07 \cdot 24 + 0.3)/0.858] + 1.6 \cdot 0.5\} = 4.03 \text{ kN/m}$$

The transverse moments are:

$$m_{d2}^- = -F_d \ (l_2/2)^2/8 = -4.03 \ (5/2)^2/8 = -3.15 \text{ kN m/m}$$
$$m_d^+ = F_d \ (l_2/2)^2/13 = 4.03 \ (5/2)^2/13 = 1.94 \text{ kN m/m}$$

The support reactions are:

$$R_1 = 1.4 \ b_e \ h_e \ 24 + (F_d/1 \text{ m}) \ (0.5 \ l_2/2) - m_d^-/(l_2/2) =$$
$$= 1.4 \cdot 0.15 \cdot 0.5 \cdot 24 + 4.03 \ (0.5 \cdot 5/2) - 3.15/(5/2) = 6.33 \text{ kN/m}$$
$$R_2 = (l_2/2) \ (F_d/1 \text{ m}) + m_d^- \ [1/(l_2/2) + 1/(l_2/2)] =$$
$$= 2.5 \cdot 4.03 + 3.15 \cdot 2/2.5 = 12.6 \text{ kN/m}$$

Checking the reactions:

$$2 R_1 + R_2 = F_d \ l_2 + 2 \ g_{e \ d}$$
$$2 \cdot 6.33 + 12.6 = 4.03 \ 5 + 2 \ 1.4 \ 0.15 \ 0.5 \ 24$$
$$25.26 \text{ kN/m} \approx 25.19 \text{ kN/m} \rightarrow \text{OK}$$

9.3.3 In-plane flexibility of plates in longitudinal direction

Let's apply the obtained reactions, R_i, in opposite directions on the upper and lower folder ribs (Figure 9.3.2a). Let's assume that N_{ij} are components of membrane forces in folder plates. An equivalent force P_i that acts in the plane of the plate, i, is a difference between the components:

$$P_i = N_{i,i-1} - N_{i-1,i} \tag{9.3.1}$$

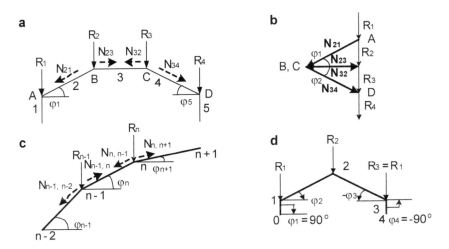

Figure 9.3.2. Considering the plate's in-plane flexibility:

a—loading the plates' structure by R_i; b—graphical method for calculating forces N_{ij};

c—membrane forces' decomposition in general case; d—a scheme for numerical example

For example, if $i = 3$ (Figure 9.3.2b)

$$P_2 = N_{21}$$

$$P_3 = N_{32} - N_{23}$$

$$P_4 = N_{34} \tag{9.3.2}$$

For a symmetric system $N_{32} = N_{23}$ and $P_3 = 0$, therefore, the expressions for membrane forces in plates (Figure 9.3.2c) are:

$$N_{n,n-1} = R_n \frac{\cos \varphi_{n+1}}{\sin (\varphi_n - \varphi_{n+1})}$$

$$N_{n,n+1} = R_n \frac{\cos \varphi_n}{\sin (\varphi_n - \varphi_{n+1})}$$

$$N_{n-1,n} = R_{n-1} \frac{\cos \varphi_{n-1}}{\sin (\varphi_{n-1} - \varphi_n)} \tag{9.3.3}$$

The equivalent force P_n that acts in the plane of plate n is an algebraic sum of the following forces (in our case it is a difference):

$$P_n = N_{n, n-1} - N_{n-1, n} = R_n \frac{\cos \varphi_{n+1}}{\sin (\varphi_n - \varphi_{n+1})} - R_{n-1} \frac{\cos \varphi_{n-1}}{\sin (\varphi_{n-1} - \varphi_n)} \tag{9.3.4}$$

9.3.4 Example of calculating the folder plates in longitudinal direction

For the folder given in Section 9.2.3 with a calculation scheme as presented in Figure 9.3.2d, it is required to calculate the equivalent membrane forces P_n ($n = 1…4$) in the plates.

Solution

Let's define the membrane forces P_1 in plate 0–1, P_2 in plate 1-2, P_3 in plate 2-3 and P_4 in plate 3–4. Then,

$$P_1 = P_4 = R_1 = 6.33 \text{ kN/m}$$

$$P_2 = P_3 = R_2 \frac{\cos \varphi_3}{\sin (\varphi_2 - \varphi_3)} - R_1 \frac{\cos \varphi_1}{\sin (\varphi_1 - \varphi_2)}$$

$$\varphi_1 = -\varphi_4 = 90°, \quad \varphi_2 = -\varphi_3 = 30.96°, \quad \varphi_2 - \varphi_3 = 61.92° = 2\varphi_2$$

$$P_2 = P_3 = 12.6 \frac{\cos (-30.96°)}{\sin 61.92°} = 12.6 \frac{\cos 30.96°}{\sin 61.92°} = 12.25 \text{ kN/m}$$

9.3.5 Bending of plates

In first approximation, it is possible to calculate the plates as simple supported beams with a span of the folder, l_1, under uniformly distributed load, P_n. The section dimensions for plates 1 and 4 (see the numerical example in the previous section) are $b_e \times h_e$ and plates 2 and 3 have dimensions $h \times [(l_2/2)/\cos \varphi_2]$. Forces P_n yield stresses $\pm \sigma_{n'}$ which are called "free edge stresses". However, there is interaction between the plate of the folder and the stresses from both sides of a rib, n, are usually different. Hence, it is required to apply the horizontal shear forces, $T_{n'}$ at each rib, n, in order to provide the stresses equilibrium from both sides of the rib.

The change in the shear forces, $T_{n'}$ corresponds to that of normal stresses, $\sigma_{n'}$ and has a parabolic form. Correspondingly, the horizontal shear stresses, $\tau_{n'}$ appear. A connection between the folder plates in horizontal direction should be designed according to these stresses.

References

Abraham, Y. 2012. Constructing one of the biggest in the world geodetic dome in the Ice Park in Eilat. Civil Engineering and Infrastructures, 50 (in Hebrew).

Billington, D. 1990. Thin shells concrete structures. McGraw-Hill publishing company, New York.

Bobrov, F.V., A.N. Byhovskiy and A.N. Gasanov. 1974. Seismic loads on Shells and Suspension Coverings, Stroyizdat, Moscow (in Russian).

EN 1992-1-1 Eurocode 2: Design of concrete structures—Part 1-1: General rules and rules for buildings, 2004.

Fisher, L. 1968. Theory and practice of shell structures. Wilhelm Ernst and Sohn, Berlin.

Housner, G.W. 1957. Interaction of building and ground during an earthquake. Bulletin of the Seismological Society of America, 47(3).

Iskhakov, I. 1981. Seismic resistant roof constructions made of RC monolithic shells with steel diaphragms. Bulletin of IASS for spatial structures.

Iskhakov, I. and G. Khaidukov. 1966. Investigations and calculation of the shallow shell models rectangular in plan with positive Gauss curvature by limit equilibrium. Beton i zelezobeton, 1 (in Russian).

Iskhakov, I. and Y. Ribakov. 2014. Collapse analysis of real RC spatial structures using known failure schemes of ferro-cement shell models. The Structural Design of Tall and Special Buildings, 23(4): 272–284.

Iskhakov, I. and Y. Ribakov. 2013. A new concept for design of fibered high strength reinforced concrete elements using ultimate limit state method. Materials and Design, 51: 612–619.

Liddel, W.I. and P.W. Miller. 1999. The design and construction of the Millennium Dome. Seminar on Space Structures, New Dehli.

Murashev, V., E. Sigalov and V. Baikov. 1971. Design of Reinforced Concrete Structures. Mir Publishers, Moscow.

Muttoni, A., F. Lurati and M.F. Ruiz. 2013. Concrete Shells—toward efficient structures: construction of an ellipsoidal shell in Switzerland, Structural Concrete, 1, 14, March.

Norfolk Scope Dome, internet resource http://hamptonroads.com/2013/11/norfolk-scope-expansion-proposed-35m-cost, accessed on April 2, 2015.

Oniashvili, O.D. 1957. Some Dynamic Problems in Theory of Shells, Moscow, USSR Academy of Science Press (in Russian).

Reisner, E. 1946. On vibration of shallow spherical shells. Journal of Applied Physics, 17(12).

Shugaev, V. and B. Sokolov. 2011. Analysis of the Reasons of Failure of Large-Span RC Roof Shells. Proceedings of IABSE-IASS Symposium, September 20–23, London.

Tomas, A. and J.P. Tovar. 2011. Imperfection sensitivity factor of the buckling load in concrete shells, Taller, longer, lighter IABSE–IASS Symposium, London.

Appendix. List of Symbols

a, b	:	half of a rectangular shell dimensions in x and y directions
b_e	:	edge element section width
D	:	dome diameter
D	:	cylindrical stiffness
c	:	width of rectangular opening in a shell
d	:	diameter of circular opening in a shell
E	:	modulus of elasticity
$E_{c\ red}$	:	reduced modulus of elasticity of concrete
E, F, G	:	coefficients of the first quadratic surface form of Gauss
G_I, G_{II}	:	first and second quadratic surface forms of Gauss
L, M, N	:	coefficients of the second quadratic surface form of Gauss
M_y	:	transverse moment in a long cylindrical shell
F_d	:	design load
f	:	rise of a shell
g	:	gravity acceleration
H	:	thrust force
H	:	height of a cylindrical shell or folder (including edge elements)
h	:	shell thickness
h_e	:	height of an edge element section
i	:	co-location point
K	:	Gaussian surface curvature
K_m	:	average curvature

k	:	curvature
k_1, k_2	:	main surface curvature at a given point
k_x, k_y, k_{xy}	:	normal curvature in directions x and y and torsion curvature
L	:	length of a curved line on a surface
l	:	shell span
M_{bc}	:	local moment
M_{pl}	:	plastic moment, carried by an RC section
M_x, M_y	:	bending moments
M_{xy}	:	torsion moment
N_1, N_2	:	internal membrane forces in a dome
N_x, N_y	:	normal membrane forces
N_{xy}	:	horizontal shear force (membrane shear)
N_{m1}, N_{m2}	:	main membrane forces
q	:	uniformly distributed load
q_{cr}	:	critical load
R	:	radius of curvature
R_1, R_2	:	main radiuses of a surface at a given point
r_0	:	radius of supporting ring of a dome
S	:	area of a shell surface
s	:	arc length
T	:	tensile force in a supporting ring of a dome
V_x, V_y	:	shear forces
w	:	vertical deflection
x, y, z	:	a system of orthogonal coordinates
z	:	internal forces lever
u, v	:	spatial coordinates on a shell surface
α_0	:	half of shell section central angle
α_1, α_2	:	inclination angles of the main forces
γ	:	density

$\varepsilon_x,\, \varepsilon_y$	:	normal membrane deformations
ε_{xy}	:	torsion deformations
v	:	Poisson coefficient of concrete
φ_0	:	half of the dome section central angle
$\varphi\,(x, y)$	:	stresses function

Index

3D model 115

T - #0443 - 071024 - C9 - 234/156/8 - PB - 9780367377212 - Gloss Lamination